A Complete Morphological Identification

Colour Atlas of Medicinal Plants

Volume - 1

Dr. Ramesh Kumar Bhutya

Former Dy. Director,
Ayurved Department,
Kota Division, Kota (Raj.)

SCIENTIFIC PUBLISHERS

Published by:
Scientific Publishers
5A, New Pali Road, P.O. Box 91
Jodhpur 342 001, India
E-mail: info@scientificpub.com
Website: www.scientificpub.com

ISBN: 978-93-83692-55-2 (Hardbound)
ISBN: 978-93-87991-87-3 (E-book)

Printed in India

Dedicated

to

Respected

Dr. Satish Kumar Sharma

R.F.S.

Assistant Conservator of Forests (Retd.)

Udaipur

Born Naturalist,

Critical Researcher, Enthusiastic Scientist

Who spent 37 years for the cause of

conservation of nature

Foreword

Medicinal plants are an important natural resource of the country which is independable for economic growth of the country and good health of our people. India is a vast country with varied forest types and agro-climatic zones. A big number of medicinal trees, shrubs, herbs, climbers, and tuberous plants grow in the country's forests and agricultural fields. China is leading in the field of herbal medicines. Our country, which is birth place of Ayurveda, is legging behind from China in this field. Our society's traditional knowledge related to medicinal plants is very vast and notable. We cannot ignore Ayurveda, medicinal plant wealth of country and our society's knowledge related to medicinal plants.

Now-a-days, awareness is seen among public for medicinal plants. Ayurveda, forest and agriculture departments are looking serious for conservation and propagation of medicinal plants. Though experts, scholars and common mass taking interest in the medicinal plants but paucity of quality literature is also felt. Correct identification of medicinal plant is a serious problem.

Dr. Ramesh Kumar Bhutya, a dedicated Ayurvedic physician, having a long experience of medicinal plants and their uses, is a prolific writer. His earlier six books namely ***'Vanaspati Aushadha Vigyan', 'Ayurvedic Medicinal Plants of India'*** (two volume), ***'Medical Ayurveda Shabdkosh', Charma Rog Nirdeshika', 'Aushadh Dravyo ka Tridosh Prabhav Vivechan', 'Anubhut Ayurveda Chikitsa'***, have been acknowledge widely. His present effort to develop a colorful atlas of medicinal plants certainly helps to identify the medicinal plants accurately. People, who have no background of plant taxonomy, they can also take help of this book to identify the medicinal plants. Definitely, this book will act as a 'Flora of medicinal plants'.

Sincerely hope, this book will be sourly liked by all the readers and research scholars.

Udaipur,
26 January, 2018

Dr. Satish Kumar Sharma
Rajasthan Forest Services (Retd.)

Prologue

Proper identification of medicinal plants is still a great problem at this time. It has also drawn the attention from common men in general and scholars of the medical streams. Wrong identification is responsible for the poor quality of the compound drugs. In the light of the above, the curiosity about the sources of the medicines and medicinal plants in most of the cases, has drawn considerable interest among all quarters of the medical science. It won't be out of place to mention here that in spite of availability of a lot of books on medicinal plants in the market, readers are not found them satisfactory as many relevant issues are not adequately addressed in them and hence, not adequately fulfill their requirements. Author has identified needs of Ayurveda people. To meet out their needs, relevant literature on medicinal plants was scanned to bring this book in present shape. Certainly, this book will help people of Ayurveda and botany in identification of medicinal plants.

In addition, in the absence of proper knowledge related to the identification of medicinal plants, many plants remained either unidentified or wrongly identified. As a result, many important and precious medicinal plants are often considered as weeds and disposed off as waste material. Wrong identification even lead towards more serious repercussions as these may be used wrongly in preparing medicines for which they are not recommended due to different chemical constituents, and hence deprived of their actual uses. In this book, an attempt is made to fill the above gaps. It is my firm belief that this book will be of immense use to Ayurveda experts, teachers, students of post-graduate and under-graduate Ayurvedic courses, researchers, students of botany, scientists, pharmacologists, pharmaceutical organizations, pharmacists, biochemists, medical personal and even common men. However, I have tried to the best of my abilities and knowledge to produce a book containing all major aspects of Ayurvedic medicinal plants, even than some gaps and laps may prevail. I will be grateful to the readers for their suggestions and valuable feedback.

Salient features of the book

1. The latest and updated botanical names are given in this book. Classical names include only those maintained in Ayurvedic literature.

2. Good photos with complete botanical identification based on characteristics of plant, stem, bark, leaf, flower, fruit, seed, tuber, root and gum with full morphological features, which would be helpful in correct identification of the plant in the field and labs.
3. Good photographs of dry material available in the market will help us to procure right material because ultimately we depend on market to have our supply of material for making the drugs. These photographs will give us a safeguard against abuse of adulteration in the dry medicinal material.
4. Photos of medicinal plants are taken in their natural habitat. All photos are original, with quality, distinguishable and taken by the author across the country.
5. At the end a composite list of Ayurvedic/Hindi names is given for readers' convenience.
6. Instead of textual botanical description the basis of identification is pictorial.

A few words of gratitude

While preparing this book, the help of various sources has been taken. I would like thank all the authorities whose works are consulted during the preparation of this book. I would like to express my grateful thanks to Dr. Satish Kumar Sharma, R.F.S., former Assistant Conservator of Forests, Udaipur, for his valuable guidance and advice. He was a constant source of inspiration throughout the creation of this book. He went out of the way to help me in spite of his very busy schedule and other professional commitments. Further, I am grateful to my colleagues Dr. R.S. Garg, Baran; Dr. P.S. Solanki, Kota; Dr. Ashok Saxena, Bundi; Dr. J.P. Sharma, Bundi; Dr. Ashutosh Sharma, Jaipur; Dr. Mohan Lal Jaiswal, Lecturer, Dravyagun Deptt., N.I.A. Jaipur; Dr. Hari Om Sharma, Ajmer; Dr. Marium Asariwala, Jodhpur; Dr. Rekha Sharma, Jaipur; Dr. Ambashankar, Sumerpur; Mr. Bajarang Lal Tailor, Bundi and Mr. Dinesh Sharma, owner of Shri Ram Herbals, Jaipur.

I thankful to Dr. D.N. Pandey, I.F.S., Member Secretary, Rajasthan Medicinal Plant Board, Jaipur, for his valuable suggestions and guidance. I am also thankful to Dr. D.S. Rajpurohit, A.C.F., Udaipur. My heartfull thanks are also due to my family members Mr. Satyawan Sharma, Mrs. Kamini Sharma, Dushyant Sharma, Mrs. Kirti Sharma, Mr. Milind and Mrs. Kumud Sharma for their constant support and assistance. Last but not the least; I am extremely grateful to my wife Mrs. Manju Sharma for her patience as she almost freed me in discharging my domestic responsibilities during writing this piece of work.

My thanks are also due to Mr. Pawan ji, Mr. Tanay ji and Rajesh Ojha, Scientific Publishers (India), Jodhpur to bring this book into existence. I am also thankful to Dr. S.L. Meena, BSI, Jodhpur for valuable suggestion and cooperation.

At last, I firmly believe this *Shloka* of *Bharatrihari Niti Shatakam*

यदा किञ्चिज्ज्ञोऽहं द्विप इव मदान्धः समभवं,
तदा सर्वज्ञोऽस्मीत्यभवदवलिप्तं मम मनः।
यदा किञ्चित्किञ्चिद् बुधजनसकाशादगतं,
तदा मूर्खोऽस्मीति ज्वर इव मदो मे व्यपगतः।।

Means: When I knew but a little, I was blinded by pride as an elephant is blinded by rut from excitement, and my mind was puffed up with the idea that I knew everything. When, however, I gradually gained knowledge through the contact of the wise I found I was a fool; and the pride, which had possessed me like fever, left me.

Thanks.

Dr. Ramesh Kumar Bhutya
3/25, Keshorai Patan,
District Bundi, Rajasthan
Mobile No.: 09414538301
E-mail: rameshbhutya@gmail.com

Contents

Abbreviations

Key characteristic for identification

H. - Habit

St. - Stem

L. - Leaf

Fl. - Flower

Fr. - Fruit

Se. - Seed

R. - Root

B. - Bulb

G. - Gum

1. *Abrus precatorius* Linn. Gunja

Family: Fabaceae Climber

Plant and Leaves

Flowers

Unripe fruits

Ripe fruits

Ripe fruits

Roots

Seeds

Key Characteristics for Identification

H. A wiry climber. **S.** Green, glabrous. **L.** Peripinnate with many pairs of leaflets, 10-20 pairs, opposite, oblong, rounded at both ends, glabrous. **Fl.** Pink, light purple, clustered **Fr.** Pods, rectangular, turgid, in bunches. **S.** Ovoid, scarlet red with a black spot, or white or black, polished.

2. *Abutilon indicum* Linn. Sweet Atibala

Family: Malvaceae Herb

Plant

Leaves

Flower

Fruits

Mericarps with seeds inside

Seeds

Key Characteristics for Identification

H. Herb or under shrub. **S.** Round, strong. **L.** Cordate, broadly ovate, acuminate, toothed, grayish green, tomentous. **Fl.** Solitary, orange-yellow or yellow, opening in evening. **Fr.** Capsule, hispid. **S.** Black or dark brown, small and flat.

3. *Acacia catechu* (Linn.f.) Willd. Khadira

Family: Mimosaceae

Tree

Tree

Leaves

Flowers

Ripe fruits

Unripe fruits

Stem

Dry stem bark

Key Characteristics for Identification

H. Tree, spiny. **S.** Dark grayish or brown rough bark and hooked short spines. **L.** 2-pinnate, polished, pinnae 10-30 pairs, leaflets 30-50 pairs. **Fl.** Small, pale yellow. **Fr.** Pods, flat, brown, shiny, 3-10 seeded, glabrous. **S.** Flat, dark brown, orbicular, rounded.

4. *Acacia nilotica* Linn.

Family: Mimosaceae

Babula

Tree

Stem Bark

Tree

Seeds

Spine

Flowers and Leaves

Leaves

Babul Gond

Fruits

Key Characteristics for Identification

H. Tree with short trunk. **S.** Erect, cylindrical, branched, solid, longitudinally deeply fissured rough bark. **L.** Oblong, entire, Pinnae 4-9 pairs, leaflets 10-25 pairs. **Fl.** Golden yellow, fragrant, globose heads. **Fr.** Lomentum, about pods with constrictions between the seeds, seeds are 8-12 per pods. ash white. **S.** Flatted, black, shiny and round.

5. *Acacia sinuta* (Lour.) Merr. Shikakai

Family: Mimosaceae

Tree

Plant

Leaves

Flowers

Ripe Fruits

Unripe Fruits

Seeds

Key Characteristics for Identification

H. Climbing shrub, armed with prickles. **S.** Branches brown. **L.** Bipinnate, sour, pinnae 4-8 pairs, 12-25 pairs. **Fl.** Small, in globose heads, yellowish-white or pink. **Fr.** Pods, flat, coriaceous, 6-10 seeded, depressed between the seeds, beaked. **S.** Round, black, shiny.

6. *Achyranthes aspera* Linn. Apamarga

Family: Amaranthaceae Herb

Plant

Stem

Leaves

Seeds

Roots

Fruits

Spikes

Key Characteristics for Identification

H. Erect, much branched, herb. **Stem:** Herbaceous, quadrangular. **Leaf:** Opposite, entire acute, rough. **Flower:** Greenish-white, spike 50 cm long, **Fruit:** Utricle, oblong-cylindrical, **Seed:** Shape like rice, single.

7. *Actinopteris dichotoma* Linn. syn.*Justicia adhatoda* Linn.

Family: Adiantaceae

Mayur Shikha

Herb

Plant

Leaves

Roots

Reverse of Leaves

Close up

Dry Plant

Key Characteristics for Identification

H. A herbaceous palm like plant. **L.** Fronds fan-like 2.5-3.8cm deep with numerous dichotomous segments which are rush-like in texture, veins few subparallel with distinct mid rib, seen on old buildings.

8. *Adansonia digitata* Forsk. Gorakh Imali

Family: Bombacaceae

Tree

Tree

Stem

Leaves (Closeup of Leaf)

Fruits (Closeup of Fruit)

Flowers (Closeup of Flower)

Seeds

Key Characteristics for Identification

H. Tree, height 10-22 m. **S.** Great diameter at base, rapidly narrowing upwards. **L.** Glabrous, leaflets, obovate-oblong, acute or obtuse. **Fl.** Solitary, white, softy, hairy. **Fr.** Like bottle-gourd, narrowed to both ends, rind hard. **S.** Shiny reniform.

9. *Adhatoda zeylanica* (Linn.) Corr. Vasa

Family: Acanthaceae

Shrub

Plant

Leaves

Flowers

Stem

Fruits

Ripe Fruits

Roots

Branch

Key Characteristics for Identification

H. Shrub or undershrub, evergreen perennial with fetid smell. **S.** Herbaceous, glabrous, erect, solid, cylindrical. **L.** Opposite, lanceolate, ovate, acute, glabrous, leathery, coriaceous. **Fl.** White, 2 purplish stripes and red spot on inner part. **Fr.** Capsule, 4 seeded, **S.** Smooth, Orbicular-oblong.

10. *Adiantum incisum* Forsk. Hansaraj

Family: Adiantaceae

Herb

Plant

Leaves

Reverse surface of leaves

Dark coloured Midrib

Key Characteristics for Identification

H. A herbaceous miniature palm like plant, **S.** Stipes tufted, wiry, spreading, dark chestnut-brown, tomentose. **L.** Fronds 15-30 cm long, simply pinnate.

11. *Aegle marmelos* (Linn.) Corr.

Family: Rutaceae

Bilva

Tree

Tree

Stem and Spines

Leaves

Flowers

Fruit pulp

Ripe Fruit

Unripe Fruit (Closeup Seeds)

Key Characteristics for Identification

H. A medium sized tree, **S.** with straight, sharp, thorns, corky bark. **L.** Trifoliate, alternate, ovate or ovate-lanceolate, crenate, acuminate. **Fl.** Greenish white, sweet scented, stamens numerous. **Fr.** Berries, 5-18 cm diameter, globose, grey or yellowish, pulp sweet yellow, aromatic.

12. *Agave americana* Linn. Kantala

Family: Agavaceae Herb

Plant

Leaves

Flower

Fruits (Close up Dry fruits)

Inflorescence

Key Characteristics for Identification

H. A succulent perennial herb. **S.** Very stout. **L.** With large rosette of 30-60 fleshy, sharply toothed. **Fl.** Clusters of yellow flowers, bloom on a pole-like stem after 10 years. **Fr.** Capsule, ovoid, globose, erect beaked,. **S.** Numerous, flattened.

13. *Ailanthus excelsa* Roxb. Aralu

Family: Simaroubaceae

Tree

Tree

Stem

Leaves

Flowers

Fruits

Key Characteristics for Identification

H. Large, tree, with bad smell. **S.** Bark is whitish, rough with irregular fissures. **L.** Pinnately compound, leaflets 8-14 pairs, ovate or falcate lanceolate, toothed sublobete. **Fl.** Yellowish. **Fr.** Samara, lanceolate, acute at both ends, reddish-brown, one seeded. **S.** Solitary.

14. *Alangium salviifolium* (Linn.f.) Wang.

Family: Alangiaceae

Ankol

Tree

Tree

Leaves

Stem

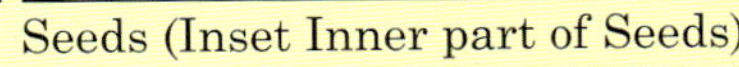

Seeds (Inset Inner part of Seeds)

Flower

Unripe Fruits (Ripe Fruits)

Bud

Key Characteristics for Identification

H. A small, scent tree. **S.** Cylindrical, dark brown, rough with shallow cracks. **L.** Alternate, narrowly oblong or ovate-lanceolate. **Fl.** Few, white or yellowish-white, fragrant. **Fr.** Berries, blue when unripe, blackish purple when ripe, sub globose, pulp sour, sweet. **S.** Oblong, brown.

15. *Albizia lebbeck* (Linn.) Willd. Shirisha

Family: Mimosaceae

Tree

Tree

Stem

Bud

Leaves

Flowering Branch

Flowers

Fruits

Key Characteristics for Identification

H. A medium to large sized, tree. **S.** Pale grey to dark brown, rough. **L.** Abruptly bipinnate, pinnae 2-4 pair, leaflets 5-9 pairs **Fl.** White, fragrant, **Fr.** Pods, linear-oblong, yellow, smooth, shiny and 4-12 seeded. **S.** Ellipsoid-oblong, compressed, pale- brown.

16. *Alectra parasitica* A. Rich.

Family: Orobancaceae

Ramanarayani Buti

Herb

Plant

Flower

Roots (Inset Dry roots)

Buds

Key Characteristics for Identification

H. This parasite herb grows on the roots of *Vitex negundo.* Herbs are 10-15 cm high, dark yellow or turmeric-like or saffron-coloured. **L.** Shape like leaves of Gram, small, crimson coloured. **Fl.** Scarlet or crimson coloured, around the catkins, petals 5.

17. *Allium cepa* Linn.

Family: Liliaceae

Palandu

Herb

Plant

Flowers and Fruits

Leaves

Seeds

Fresh Tuber

Tuber

Key Characteristics for Identification

H. Herb with aromatic bulb. **S.** Underground. **L.** In two rows, linear, acute, hollow, cylindrical, flashy, glaucous green. **Fl.** Many, greenish-white. **Fr.** Capsule, globose, membranous. **S.** Black, small, compressed, 3-angular, shiny. **B.** Thick, globose, tunicate, flashy, odour characteristic.

18. *Allium sativum* Linn. Rasona

Family: Liliaceae Herb

Plant

Leaves

Tuber

Flowers and Fruits

Key Characteristics for Identification

H. A foetid herb, with bulbs. **B.** Consisted of smaller bulbs called cloves which are 8-12 in numbers, bulbs covered over by outer white, thin scales. **S.** Smooth, round. **L.** Long simple, linear, flat, acute. **Fl.** Small and white. **Fr.** Capsule small.

19. *Aloe barbadensis* Mill. Ghrita kumari

Family: Liliaceae

Herb

Plant

Leaves (Inset Extract of leaves)

Fruits

Flowers (Inset inner part of flower)

Seeds

Key Characteristics for Identification

H. A succulent, coarse herb. **S.** Short, thick. **L.** Fleshy, sessile, glaucous-green with white blotches, crowded, lenceolate, erect-spreading, rather concave, convex below, 45-60cm long, 5-10cm wide. **Fl.** Yellow or red or orange. **Fr.** Loculicidal capsule, cylindrical or ellipsoid-oblong.

20. *Alstonia scholaris* R.Br. Saptaparna

Family: Apocynaceae

Tree

Tree

Stem

Flower

Leaves

Fruits

Key Characteristics for Identification

H. A evergreen tree, with milky latex. **S.** Straight bole, bark grayish-brown. **L.** 4-7 in a whorl, usually 6, glabrous, coriaceous, oblong-lanceolate or obovate. **Fl.** Greenish-white, small, strongly fragrant. **Fr.** Follicles, 30-60 cm long, in pairs or clusters, cylindrical. **S.** White, flattened.

21. *Annona squamosa* Linn. Sitaphala

Family: Annonaceae Tree

Leaves

Tree Seeds

Stem Flower Fruit

Key Characteristics for Identification

H. A small, evergreen tree. **S.** Herbaceous, cylindrical. **L.** Ovate, entire, acute, glabrous above, oblong-lanciolate or elliptic-obtuse. **Fl.** Yellowish-green, solitary. **Fr.** Ataerio of berries, yellowish-green, fruit pulp white, prominent glands on dorsal surface. **S.** Hard, brownish-black shiny.

22. *Anogeissus latifolia* Wall. ex Bedd.

Family: Combrataceae

Dhava

Tree

Tree

Stem

Leaves (Inset Lower surface of leaves)

Fruits

Flowers

Flower Branch

Gum

Key Characteristics for Identification

H. A large tree. **S.** Greenish- white, smooth bark. **L.** Alternate or sub-opposite, elliptic or oblong-elliptic, glabrous, leaves turning red before falling. **Fl.** Sessile, minute, greenish-white. **Fr.** yellowish-brown, small, shining, beaked, with persistent calyx. **S.** Wedge shaped, solitary.

23. *Anthocephalus chinensis* A Rich. ex Walp.

Family: Rubiaceae

Kadamba

Tree

Tree

Stem

Leaves

Fruits

Flowers

Raw Wood

Key Characteristics for Identification

H. An evergreen tree. **S.** Bark dark brown, rough. **L.** Elliptic-oblong or ovate, glabrous. **Fl.** Small, orange or yellow in terminal, solitary globose heads. **Fr.** A globose pseudocarp, green when unripe, orange or yellow when ripe, fleshy, sweet-sour. **S.** Small, muriculate.

24. *Argemone mexicana* Linn. Satyanashi

Family: Papaveraceae

Herb

Leaves

Plant

Fruits

Flower

Ripe Fruit

Seeds

Key Characteristics for Identification

H. A prickly or spiny, Herb, with yellow latex. **S.** Herbaceous, cylindrical, branched, solid, spiny, green. **L.** Bristle with slightly incised and undulated margins, acute, surface spiny. **Fl.** Large, bright yellow bowl shaped. **Fr.** Capsule, surface prickly, oblong, ovoid. **S.** Numerous, brown.

25. *Argyreia nervosa* (Burm.f.) Boj.

Family: Convolvulaceae

Vidhayara

Climber

Plant

Leaves (Closeup of Reverse Leaf)

Stem

Flowers

Fruits (Closeup of Fruits)

Seeds

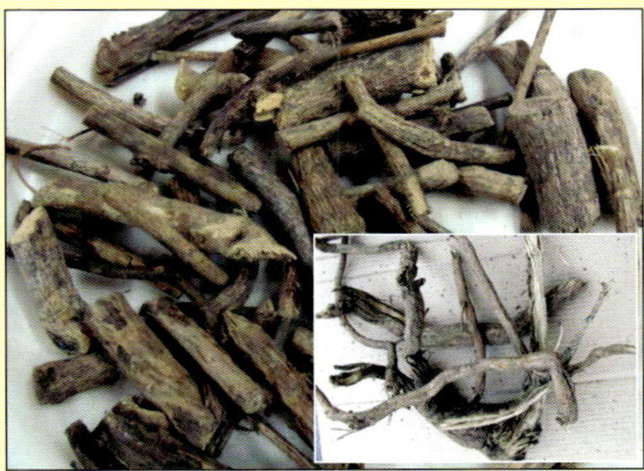

Raw Roots (Inset fresh roots)

Key Characteristics for Identification

H. A large woody climber. **S.** White, tomentose. **L.** Large, Ovate-cordate, acute, glabrous above, white dense tomentose of soft silky beneath. **Fl.** Large, bell shaped, rose-purple, open in night, fragrant. **Fr.** Globose, green when unripe, brown yellow when ripe. **S.** Triangular, whitish-brown.

26. *Aristolochia bracteolata* Lam. Keetamari

Family: Aristolochiaceae

Herb

Plant

Fruit

Flower

Reverse part of Leaves

Seeds

Key Characteristics for Identification

H. A slender herb. **S.** Week, glabrous, prostrate, branched. **L.** Alternate, reniform or broadly ovate. **Fl.** Solitary, with a large, sessile orbicular, bract at the base, trumpet shaped mouth, lip linear, dark purple. **Fr.** Capsule, oblong-ellipsoid, 12 ribbed, glabrous. **S.** deltoid, black.

27. *Aristolochia indica* Linn. Ishwari

Family: Aristolochiaceae Climber

Plant

Leaves

Fruits

Dry Fruits

Closeup of Fruits

Key Characteristics for Identification

H. A glabrous climbers. **S.** Greenish or pale. **L.** Simple, alternate, usually obovate-oblong to sub-pandurate, somewhat cordate, acuminate. **Fl.** Pale-green or light purplish, funnel-shaped. **Fr.** Capsule, oblong or globose-oblong, hexa-gonal. **S.** Flat, brown, winged, deltoid-ovate, acute.

28. *Artocarpus heterophyllus* Lam. Katahala

Family: Moraceae

Tree

Tree

Stem

Leaves

Fruits

Closeup of Fruits

Bud

Key Characteristics for Identification

H. A large, evergreen tree, rounded crown. **S.** Bark black mottled. **L.** Alternate, shiny, green above. **Fl.** Male flowers crowned on cylindrical receptacles, female flowers crowned on globose receptacles, yellowish. **Fr.** Multiple, large, fleshy, globose or oblong. **S.** Oval, reniform.

29. *Asparagus racemosus* Willd.

Family: Asparagaceae

Shatavari

Climber

Plant and Leaves

Flowers

Fruits

Roots

Unripe Fruits

Dry Seeds

Roots with Plants

Raw Roots

Key Characteristics for Identification

H. Climbing under shrub, ornamental, prickly. **S.** Aerial, with recurved spines, much branched. **L.** Alternate, reduced to minute chaffy scales and spines. **Fl.** Bracteate, white, fragrant. **Fr.** Berry, globular or obscurely 3-lobed, red or purplish black when ripe, 3 seeded. **S.** Black.

30. *Azadirachta indica* A. Juss. Neem

Family: Meliaceae

Tree

Tree

Stem

Flowers

Seeds

Fruits

Leaves with inflorance

Key Characteristics for Identification

H. A large, evergreen tree. **S.** Dark grey tubercled bark, channeled, tough, fibrous, rough scaly surface. **L.** Alternate, crowded at the end of branches, unequal sides, dark green, glabrous, ovate-lanceolate. **Fl.** White, small. **Fr.** Drupe, scented, green, turning yellow on ripping, oblong. **S.** Ellipsoid, cotyledons thick, fleshy and oily.

31. *Bacopa monnieri* (Linn.) Penn. Brahmi

Family: Scrophulariaceae

Herb

Plant (Inset Raw Plant)

Leaves and Flowers

Flower

Key Characteristics for Identification

H. A glabrous, prostrate, juicy, herb. **S.** Rooting at the nodes, branches numerous. **Leaf:** Simple, opposite, decussate, obovate-oblong or spathulate, rather fleshy. **Fl.** Auxiliary, solitary, pale-blue or whitish, on long slender pedicles. **Fr.** Ovoid, acute, 2-celled capsule. **S.** Minute, numerous.

32. *Balanites aegyptiaca* (Linn.) Delile

Family: Balanitaceae

Hingot

Tree

Tree Stem Flowers

Fruits

Ripe, unripe fruits and seed

Leaves

Unripe Fruits (Inset inner part of Seeds)

Key Characteristics for Identification

H. A small, spiny tree. **S.** Twigs armed with stout axillary. **L.** Alternate, bifoliate, leaflets, elliptic-obovate, oblanceolate. **Fl.** Greenish-white, fragrant, 4-12 in a group. **Fr.** Drupes, 3-6 cm long & 2-4 cm wide, ovoid, woody, pulpy, 5 grooved, pale-yellow when ripe. **S.** Solitary, oily.

33. *Bambusa arundinacea* (Retz.) Willd.

Family: Bambusaceae

Bansa

Tree

Plant

Leaves

Stem Flowers Gum Synthetic Gum

Key Characteristics for Identification

H. A tall thorny tree, with many stems. **S.** Graceful and curving, nodes prominent, stem hollow between joints. **L.** Linear or linear-lanceolate. **Flower:** Yellow or yellowish-green, in very long panicles. **Fruit:** Oblong, grains 5-8.3mm long, beaked by the style base, grooved on one side.

34. *Barleria prionitis* Linn. Saireyaka

Family: Acanthaceae

Herb

Plant

Leaves

Roots

Flower

Leaves and Flower

Fruit

Key Characteristics for Identification

H. An prickly much branched, herb or under shrub. **S.** Herbaceous, erect, angular, branched, solid, green, 4-gonous, glabrous. **L.** Elliptic, acuminate, bristle-tipped, entire, lineolate, glabrous above. **Fl.** Sessile, complete, pentamerous, yellow. **Fr.** Capsule, 2 seeded, ovoid. **S.** Compressed.

35. *Barringtonia acutangula* (Linn.) Gaertn.

Family: Barringtoniaceae

Samudraphala

Tree

Plant

Leaves

Fruits

Ripe Fruits

Inflorance

Key Characteristics for Identification

H. A medium sized tree. **S.** Rough, dark brown bark. **L.** Alternate, obovate-oblong or elliptic. **Fl.** Fragrant, dark scarlet in pendulous, inflorescence 30-60 cm long. **Fr.** Drupe, green, oblong, bluntly quadrangular, narrowed towards the ends, shape like greater cardamom **S.** Ovoid, black.

36. *Bauhinia variegata* Lamk. Kachanar

Family: Ceasalpiniaceae Tree

Tree

Stem

Leaves

Root Stem Bark

Flowers

Unripe Fruits

Fruits

Key Characteristics for Identification

H. A shrub or tree. **S.** Bark dark scabrous, branches numerous. **L.** Broader than long, 2 rounded lobes, green and glabrous. **Fl.** White or pale yellow. **Fr.** Pods, stalked, 12-25 by 1.3-2.2 cm, blunt at the apex, tapering to the base, glabrous, black on ripe. **S.** 12-20, oblong, black, compressed.

37. *Benincasa hispida* (Thunb.) Cogn.

Family: Cucurbitaceae

Kushamanda

Climber

Flower and Leaves

Fruits

Seeds

Leaves

Key Characteristics for Identification

H. A climbing gourd. **S.** Angular, softly hairy, climbing by tendrils. **L.** Large, cordate, reniform-orbicular, deeply 5-lobed. **Fl.** Large, yellow. **Fr.** large, fleshy, oblong, broadly cylindric, pubescent not ribbed, hairy, bluish-white, fruit-pulp white. **S.** Oblong, compressed, yellowish-white, oily.

38. *Bixa orellana* Linn.

Family: Bixaceae

Sinduri

Tree

Plant

Leaves

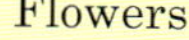

Flowers

Fruit

Unripe Fruit

Seeds

Key Characteristics for Identification

H. A small evergreen tree. **S.** Erect, branched, solid, smooth and grayish-green. **L.** Large, ovate, sub cordate at base, acuminate, glabrous on both surface. **Fl.** White or pink. **Fr.** Capsule ovoid or sub globuse, reddish brown or green. **S.** Trigonous, rounded, covered with a red pulp.

39. *Boerhavia diffusa* Linn. Punarnava

Family: Nyctaginaceae

Herb

Plant

Leaves

Flower

Roots

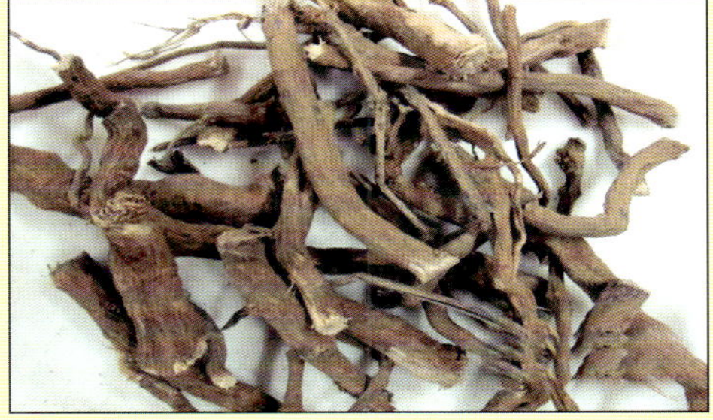
Dry Roots

Key Characteristics for Identification

H. A prostrate herb. **R.** Fusiform, twisted. **S.** Herbaceous, reddish, glabrous. **L.** Opposite, as broad as long, ovate or oblong-cordate, glabrous above, and whitish beneath. **Fl.** Red, pink or white, small, 4-10 in number. **Fr.** Ovate, oblong, one-seeded, 5-ribbed. **S.** Small, black, reniform.

40. *Bombax ceiba* Linn.

Semal

Family: Bombacaceae

Tree

Tree

Stem and Fruits

Leaves

Leafless plant and spines

Flower

Buds

Semal Beej

Gum

Young Roots (Inset Dry young root's pieces)

Key Characteristics for Identification

H. A tall tree. **S.** Bark grey or brown covered with black, hard, conical prickles. **L.** Digitate, large, spreading, crowed at the ends of branches; leaflets 5-7, elliptic-lanceolate, glabrous. **Fl.** Bright red, numerous, crowed at the end of branches. **Fr.** Capsules, ellipsoid, 5-angled. **S.** Ovoid, black or grey.

41. *Boswellia serrata* Roxb. Shallaki

Family: Bursaraceae

Tree

Tree Leafless Tree Stem

Leaves

Flowers

Fruits

Gum

Key Characteristics for Identification

H. A medium to large tree. **S.** Ash coloured papery bark peeling off in thin flakes. **L.** Imparipinnate, alternate, leaflets 8-16 pairs, opposite, ovate or ovate-lanceolate, crenate-serrate. **Fl.** Small, fragrant, white. **Fr.** Drups, trigonus, 3-valved, scarlet red. **S.** Compressed.

42. *Brassica campestris* Linn. Sarson

Family: Brassicaceae

Herb

Plants

Leaves

Fruits

Flower

Sarson

Key Characteristics for Identification

H. Herb, glaucous. **S.** Herbaceous, cylindrical, branched, solid, smooth and green. **L.** Alternate, basal leaves are 25 cm long, lyrate, pinnatified, dentate and broad, with deeply cut margins, upper leaves are 12 cm long, dentate. **Fl.** Yellow, attractive. **Fr.** Siliqua, beaked. **S.** Yellow, round.

43. *Buchanania lanzan* Spreng. Chironji

Family: Anacardiaceae

Tree

Tree (Inset Stem)

Leaves

Flowers

Ripe Fruits

Seeds

Unripe Fruits

Key Characteristics for Identification

H. A evergreen medium sized tree. **S.** Thick, dark grey or nearly black rough fissured bark. **L.** Thickly coriaceous, oblong, obtuse. **Fl.** Small, greenish white. **Fr.** Drupes, ovoid or globose, green when unripe, turns black when ripe. **S.** Hard, 2-valved, reddish brown, fragrant and oily.

44. *Butea monosperma* (Lam.) Taub.

Family: Papilionaceae

Palasha

Tree

Plant (Inset Stem)

Leaves

Flowers (Inset dry flowers)

Yellow Flowers with Fruits

Fruits

Seeds

Dry Root Bark

Gum

Key Characteristics for Identification

H. A medium sized tree. **S.** Ash grey bark, crooked trunk. **L.** 3-foliate, leaflets coriaceous, broadly obovate with cuneate or deltoid base, obtuse, glabrous. **Fl.** Bright orange red, large, 3 flowers together. **Fr.** Pods silky, tomentose, containing a single seed. **S.** Flat, reniform.

45. *Caesalpinia bonduc* (L.) Roxb. Dandy & Exell.

Family: Ceasalpiniaceae

Kantaki Karanja

Climber

Plant

Leaves

Flower

Fruits

Dry Fruits

Open Fruit

Spines

Seeds

Key Characteristics for Identification

H. An thorny shrub or climber. **S.** Armed with hooks and straight hard yellow prickles. **L.** Bipinnate, pinnae 6-11 pairs, leaflets 6-9 pairs, elliptic-oblong, obtuse, strongly mucronate, glabrous above. **Fl.** Yellow, in dense long racemes. **Fr.:** Pods, oblong. **S.** 1-2 per pod, oblong , hard.

46. *Cajanus cajan* (Linn.) Mill. Arahara

Family: Fabaceae

Shrub

Plant

Leaves

Flower Buds

Flowers

Fruits

Seeds

Key Characteristics for Identification

H. Shrub, erect. **S.** Herbaceous, lower portion woody, soft hairy. **L.** Alternate, trifoliate compound, leaf base pulvinous, stipulate, unicostate reticulate. **Fl.** Yellow, pentamerous. **Fr.** Legume, 5-7.5 cm long, silky, purple spotted, 3-4 seeded. **S.** Shape like Pea, vary in colour.

47. *Calotropis procera* (Willd.) Arka

Family: Asclepiadaceae

Shrub

Plant (Inset Stem)

Leaves & Flowers

Young Branch

Fruits

Seeds

Dry roots

Key Characteristics for Identification

H. A milky shrub, all parts of shrub are covered with loose soft white wool. **S.** Herbaceous, bark soft, corky, spongy. **L.** Covered with soft white wool, opposite, ovate, ovate-oblong. **Fl.** Lilac, white, rosy or purple. **Fr.** Ettaerio of follicle, fleshy, green. **S.** Ovate, acute, flattened black.

48. *Capparis deciduas* (Forsk.) Edgew

Family: Capparidaceae

Karira

Tree

Flowers

Tree (Inset Stem)

Flowers (Yellow)

Unripe Fruits

Fruits

Seeds

Key Characteristics for Identification

H. Shrub or small tree, spiny. **S.** Woody, rough, dense branched, smooth and green with spines. **L.** Simple, linear, oblong, alternate, stipular thorns long, sharp, straight, orange yellow. **Fl.** Reddish pink or yellow. **Fr.** Berry, globular, glabrous, green, red or pink when ripe. **S** White, small.

49. *Capsicum annum* Linn. Lal Mircha

Family: Solanaceae

Herb

Plant

Unripe Fruit

Flower

Leaves

Seeds

Key Characteristics for Identification

H. A herb or shrub. **S.** Angular, branches also angular. **L.** Simple, ovate, entire, acuminate, usually wrinkled, dark green. **Fl.** White or violet, in cluster two or more. **Fr.** Long, cylindrical, ovoid, obtuse or oblong, green when unripe and bright red when ripe. **S.** Yellow, smooth, round.

50. *Carica papaya* Linn. Eranda Karkati

Family: Caricaceae

Tree

Tree (Inset Stem)

Leaves

Male Flowers

Fruits

Female Flowers

Seeds

Key Characteristics for Identification

H. A soft wooded, lacticiferous tree. **S.** Non-branched, straight, cylindrical, with a tuft of leaves at the top, with latex. **L.** Large, palmately, 7-lobed, each lobe is pinnately lobed again,. **Fl.** Greenish-yellow. **Fr.** Berries, fleshy, ovoid-oblong, pulp yellow. **S.** Many, spherical, black shiny.

51. *Carissa carandas* Linn. Karonda

Family: Apocynaceae Shrub

Plant (Inset Stem)

Leaves

Flowers

Fruits (Inset Closeup of Fruits)

Key Characteristics for Identification

H. A spiny, evergreen shrub. **S.** With short stem, branches 2-3 chotomous, spines strong, bark light grey. **L.** Simple, opposite, shining, elliptic-oblong or oblong. **Fl.** White, fragrant. **Fr.** Berries, ellipsoid or globose, green when unripe, purplish-black, smooth, very sour. **S.** 4-8, usually 2.

52. *Cassia alata* Linn. Dadmaran

Family: Ceasalpiniaceae Herb

Plant

Leaves

Flowers (Inset closeup flower)

Fruits

Seeds

Key Characteristics for Identification

H. A large shrub. **S.** Very thick finely downy branches. **L.** Subsessile, 30-60cm long, leaflets 8-12 pairs, oblong-obtuse, glabrous. **Fl.** In short pedicles, pendiculated racemes, bright yellow with darker veins, broad-ovate. **Fr.** Pods long, ligulate with a broad wing, membranous. **S.** 50 or more.

53. *Cassia auriculata* Linn. Charmaranga

Family: Ceasalpiniaceae

Shrub

Plant

Leaves

Bud

Key Characteristics for Identification

H. A shrub. **S.** Bark smooth, with reddish-brown branches. **L.** Leaflets oblong-obovate, obtuse or, glabrous, dull green above, paler beneath. **Fl.** Large, bright yellow. **Fr.** Pods, flat, thin, oblong, obtuse, pale brown, depressed between the seeds. **S.** 10-20 per pods, obovate,

54. *Cassia fistula* Linn. Amalatasa

Family: Ceasalpiniaceae

Tree

Tree

Leaves

Flowers

Fruits (Closeup of Fruits)

Plant at Flowering

Unripe Fruits

Dry Fruit Pulp

Seeds

Key Characteristics for Identification

H. A moderate sized tree. **S.** Smooth and greenish-gray. **L.** Compound, leaflets 4-8 pairs, alternate, ovate, acute, bright green. **Fl.** Bright yellow and fragrant. **Fr.** Legume nearly straight, smooth, shiny, brownish black. **S.** Broadly ovate, flat, reddish-brown, in dark coloured sweetish pulp.

55. *Cassia occidentalis* Linn. Kasani

Family: Ceasalpiniaceae Herb

Plant

Leaves

Flowers and Fruits

Seeds

Key Characteristics for Identification

H. Under shrub, odorous. **S.** Herbaceous, glabrous, branches red. **L.** Alternate, compound, leaflets 4-5 pairs, elliptic-lanciolate. **Fl.** Yellow, pentamerous. **Fr.** Legume, cylindrical or compressed, glabrous pods, containing 20-30 seeds. **S.** Ovoid, compressed, hard, shiny, green.

56. *Cassia senna* Linn.

Sanaya

Family: Ceasalpiniaceae

Herb

Plant & Leaves

Ripe Fruits

Flowers

Raw Leaves

Unripe Fruits

Seeds

Key Characteristics for Identification

H. A shrub or undershrub. **S.** With pale obtusely angled branches. **L.** Paripinnate, leaflets 3-9 pairs, pale green oblong or ovate, glabrous. **Fl.** Brilliant yellow, in axillary racemes. **Fr.** Pods, flat, thin, oblong, light green when young to dark brown when ripe. **S.** Smooth, brown, obovate-oblong.

57. *Cassia tora* Linn. Chakramarda

Family: Ceasalpiniaceae Herb

Plant

Leaves (Inset Slipping action at evening)

Flowers

Fruits

Leaves and Flowers

Seeds

Key Characteristics for Identification

H. A foetid herb or undershrub. **S.** Cylindrical, branched. **L.** Compound, paripinnate, leaflets 3 pairs, joins at night, opposite, obovate-oblong, glabrous, typical smell. **Fl.** Bright yellow, 1.25 cm across, in sub-sessile pairs. **Fr.** Pods long, slender. **S.** Rhombohedral, dark brown, very hard.

58. *Catharanthus roseus*

(L.) G. Don.

Family: Apocynaceae

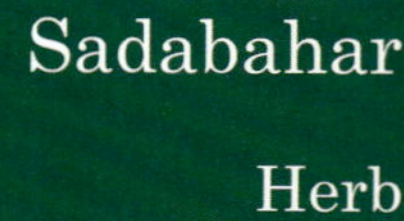

Sadabahar

Herb

Plant

Leaves

Flower Red

Flower White

Key Characteristics for Identification

H. A handsome herb. **S.** Herbaceous, angular, branched, solid, purple-red, milky latex present. **L.** Dark green, oval, oblong or obovate, glossy, opposite, decussate, stipulate, interpetiolar, simple, latex present. **Fl.** White or deep pink, petals 5. **Fr.** Etaerio of follicles. **S.** Blackish-brown.

59. *Celastrus paniculatus* Willd.

Jyotishamati

Climber

Family: Celastraceae

Plant

Leaves (Inset Close up of Leaves)

Fruiting Plant

Fruits

Fresh Seeds (Inset Closeup of Seeds)

Flowers

Lenticles preasent on stem

Dry Seeds

Key Characteristics for Identification

H. A large, woody, climbing shrub. **S.** With long branches which are reddish-brown and covered with white lenticles. **L.** Simple, alternate, broadly elliptic-ovate or obovate, glabrous. **Fl.** Greenish-white. **Fr.** Capsules, bright yellow when ripe. **S.** Ellipsoid or ovoid, yellowish or reddish-brown.

60. *Centella asiatica* (Linn.) Urban.

Mandukaparni

Family: Apiaceae

Herb

Plant-carpet of plant growing with grasses

Fruits

Leaves (Inset Closeup of Leaves)

Key Characteristics for Identification

H. A herbaceous creeping herb, rooting at the nodes. **S.** Long, prostrate, reddish, and with long internodes. **L.** Simple, cordate, orbicular reniform, glabrous on both sides, toothed. **Fl.** Pink, red or white. **Fr.** Oblong, dull brown, ovoid, hard, crowned by the persistent petals. **S.** Flat, small.

61. *Centipeda orbicularis* Lour.

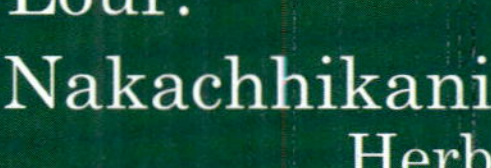

Nakachhikani

Family: Asteraceae

Herb

Plant

Fruit

Leaves

Fruits

Key Characteristics for Identification

H. A prostrate herb. **S.** Numerous, 10-20 cm long, glabrous. **L.** Numerous, oblong-spathulate, with few teeth, usually 2 on each side, shape like leaves of Carrot but small. **Fl.** Yellow or white, solitary, axillary, sub sessile. **Fr.** Achenes, minute, 4-angled. **S.** Small, round, black.

62. *Chlorophytum tuberosum* Baker.

Mushali swet

Family: Liliaceae

Herb

Plant

Leaves

Flowers (Inset Flower)

Fruits

Dry Roots

Root

Seeds

Key Characteristics for Identification

H. A low herb. **S.** Root fibers ending into ellipsoid tubers. **L.** Subdistichous, sessile, strongly recurved, grass like or conduplicate or flat, linear-lanceolate, falcate. **Fl.** Solitary or paired, pedicels 2-5mm, bracts deltoid. **Fr.** Capsule obcordate, 3-winged. **S.** Angular, 1 in each valve.

63. *Cicer arietinum* Linn. Chana

Family: Papilionaceae

Herb

Plant

Leaves flower surface

Flowers

Fruits

Seeds

Key Characteristics for Identification

H. An erect or spreading, herb. **S.** Much and spreading branched, covered over with glandular hairs. **L.** Imparipinnately compound, leaflets 9-15 pairs, ovate, oblong or obovate. **Fl.** Pink, blue or white. **Fr.** Ovoid to trapezoid, often 2-seeded. **S.** Obovate, shortly beaked, reddish brown.

64. *Cichorium intybus* Linn. Kasani

Family: Asteraceae Herb

Plant

Leaves

Flower

Root

Key Characteristics for Identification

H. A glandular milky herb. **S.** Herbaceous, with tough, rigid, spreading branches. **L.** Redical and lower pinnatified, lobes toothed, upper alternate, small, entire. **Fl.** Bright blue, in ligulate heads, terminal. **Fr.** Achenes, smooth, 5-angled, tubernate, crowned with a ring of erect pappus scales.

65. *Cissampelos pareira* Linn. Patha

Family: Menispermaceae Climber

Plant

Leaves

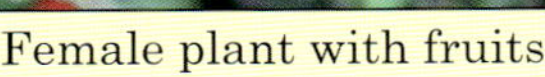

Female plant with fruits

Flower and Lower surface of leaves

Key Characteristics for Identification

H. A climber shrub. **S.** Branches striate, pubescent or sub glabrous. **L.** Peltate or orbicular-reniform or ovate-sub reniform, pubescent on both surface. **Fl.** Minute, greenish-yellow. **Fr.** Drupes, small, ovoid-sub globose or obovoid, compressed, scarlet red. **S.** Horse-shoe shaped.

66. *Cissus quandrangularis* Linn.

Family: Vitaceae

Hadajora

Climber

Plant

Stem and Tendrils

Leaves

Fruits

Flowers

Key Characteristics for Identification

H. A tendril climber. **S.** Stout, fleshy, jointed, slender, glabrous, much contracted at the nodes, quandrangular. **L.** Ovate or reniform, some times 3-7 lobed, cordate, glabrous, fleshy. **Fl.** Small, greenish-white. **Fr.** Berry, obovoid or globose, shiny, red when ripe. **S.** Ellipsoid, dark brown.

67. *Citrullus colocynthis* (L.) Schard.

Family: Cucurbitaceae

Indrayana

Climber

Plant

Leaves

Flower

Fruit (Closeup Fruit)

Seed

Seeds

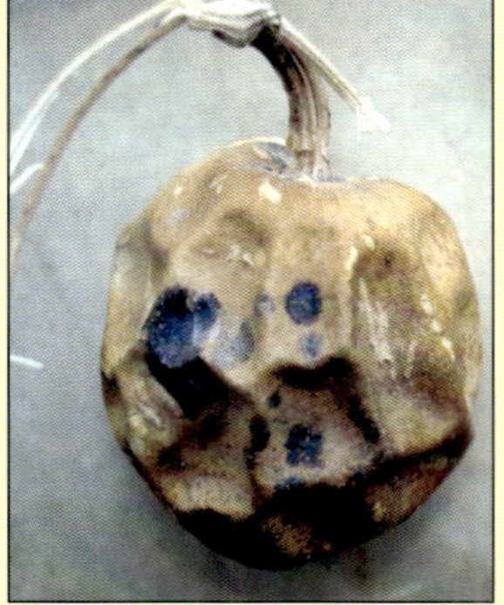

Dry Fruit

Ripe Fruit

Key Characteristics for Identification

H. A trailing herb or climber. **S.** Diffuse or creeping, angled, hirsute and bifid hairy tendrils. **L.** 3-7 lobed, lobe deeply pinnatifid. **Fl.** Pale-yellow. **Fr.** Berry, globose, green when unripe and pale yellow when ripe, glabrous when ripe, 10 longitudinal ribs, red in colour. **S.** Numerous, ovoid-oblong.

68. *Citrullus vulgaris* Schard. Tarbooja

Family: Cucurbitaceae

Climber

Plant

Leaves

Fruits

Cross section of fruit

Seeds

Inner part of seeds

Key Characteristics for Identification

H. A extensively climber. **S.** Very long, 10-15m, thick, angular, endrils bifid, stout, pubescent. **L.** Deeply divided, but moderately lobed, glabrous. **Fl.** Yellow, greenish. **Fr.** Sub globose or ellipsoid, smooth, greenish or clouded, flesh juicy, red, sweet. **S.** Black or brown, shiny and flat.

69. *Citrus aurantifolia* (Christm) Swimgle

Nimbu

Shrub

Family: Rutaceae

Plant

Leaves

Flower

Ripe Fruits

Fruits

Seeds

Key Characteristics for Identification

H. A thorny shrub or tree. **S.** With greenish-white, glabrous. **L.** Foliolate, leaf stalks broadly winged, elliptic or ovate. **Fl.** Bisexual, pure white, large, very fragrant. **Fr.** Globose, green when unripe and bright yellow when ripe, pulp sour. **S.** Numerous, creamy coloured, smooth, slimy.

70. *Cleome gynandra* Linn. Hulhula

Family: Capparidaceae

Herb

Plant

Leaves

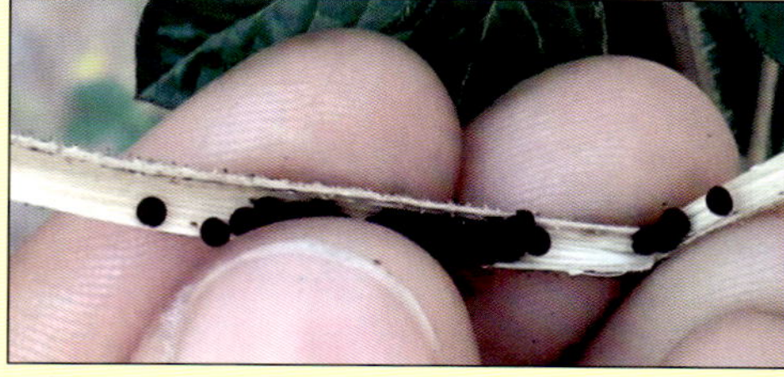

Flower

Seeds

Fruits

Key Characteristics for Identification

H. A sticky herb. **S.** Aerial, herbaceous, hairy and green. **L.** 3-5 foliate, palmately compound. **Fl.** White, yellow or purple as species, ebractiate, pedicilate, tetramerous. **Fr.** Capsule, long, round, stalk long, 5-7.5 cm long, acute, striped, glandular, hairy and compressed. **S.** Brownish-black.

71. *Clitoria ternatea* Linn. Aparajita

Family: Fabaceae

Herb

Plant

Fruits and Leaves

Flower White

Flower Blue

Seeds

Key Characteristics for Identification

H. A twining climber or herb. **S.** Slender, terete, downy branches. **L.** Compound, imperipinnate, leaflets 5-7, elliptic-oblong, obtuse, glabrous. **Fl.** Blue or white, axillary, solitary, pedicels. **Fr.** Pods, flattened, sharp beaked. **S.** Black or yellowish-brown, smooth, ovate, flattened.

72. *Cocculus hirsutus* (Linn.) Diels

Family: Menispermaceae

Pataalgarudi

Climber

Plant

Leaves

Back of Leaves

Fruits

Seeds

Key Characteristics for Identification

H. A straggling shrub. **S.** Young parts densely and softly villous. **L.** simple, alternate, ovate-oblong, obtuse, apiculate, softly villous, nearly glabrous. **Fl.** Male flowers small, in axillary cymose penicles, female flowers 2-3 together in axillary clusters. **Fr.** Drupes, purplish black

73. *Cochlospermum gossypium* D.C.

Family: Cochlospermaceae

Katira
Tree

Leaves

Flowers

Fruit

Gum

Key Characteristics for Identification

H. A small tree. **S.** Ash coloured, smooth, deeply furrowed bark. **L.** Scattered about the ends of the branchlets, palmately 3-5 lobed, glabrous above. **Fl.** Large, golden yellow. **Fr.** Capsules, large, obovoid, striate outside, pear shaped, 3-5 valved. **S.** Cocheate, with white silky woolly hairs.

74. *Cocos nucifera* Linn. Nariyal

Family: Aracaceae Tree

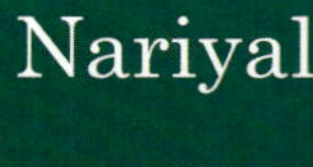

Plant

Stem

Unripe Fruit

Leaves

Fruits (Inset: Close up of Fruits)

Key Characteristics for Identification

H. A non-branched stately tree. **S.** Surrounded by a mass of rootlets, cylindrical, with ring like leaf-scars. **L.** Leaflets narrow and tapering, linear-lanceolate. **Fl.** Spadix 1.2-1.8m long, yellow or orange flowers. **Fr.** 3-gonously obovoid or sub-globose, green or yellowish, drupes. **S.** One, oval.

75. *Coix lacryma-jobi* Linn. Gavedhuk

Family: Poaceae Herb

Plant (Inset Closeup of Flower)

Leaves and Fruits Flowers

Key Characteristics for Identification

H. A erect grass herb. **S.** With basal rooting nodes, stout, internodes smooth, polished. **L.** Linear-lanceolate, smooth, membranous. **Fl.** Racemes 2.5-6.3 cm long, carrying male and female spikelets. **Fr.** Sub-globose or ellipsoid, bluish grey, smooth, polished, tear shaped grains.

76. *Commiphora agallocha* Linn.

Family: Burseraceae

Guggulu badi

Shrub

Plant (Inset Gum)

Leaves

Stem

Spines

Key Characteristics for Identification

H. A small crooked tree. **S.** Bark ash coloured, rough, peeling of in flakes, pubescent, branches crooked and knotty. **L.** Alternate, oval or elliptic, shining. **Fl.** Red, small. **Fr.** Drupes, red, smooth, the size of a black currant, nut 2-celled, with a single seed. **G.** Pale yellow or brown, aromatic.

77. *Commiphora mukul* (Arn.) Bhandari

Family: Burseraceae

Guggulu

Shrub

Plant

Stem

Leaves

Gum

Buds

Gum Production

Key Characteristics for Identification

H. An armed shrub or tree. **S.** Bark ash coloured, rough, peeling of in flakes, young parts glandular, pubescent, branches crooked and knotty. **L.** Palmately trifoliate, alternate, shining. **Fl.** Small, brownish-red. **Fr.** Drupes, red when ripe, ovoid. **G.** Pale yellow or brown, aromatic.

78. *Convolvulus pluricaulis* Choisy.

Family: Convolvulaceae

Shankhapushpi

Herb

Plant (Inset Photo : Plants)

Leaves

Flowers (Inset Close up of Flower)

Key Characteristics for Identification

H. A trailing prostrate or sub erect, spreading hairy herb, 10-30 cm long. **S.** Herbaceous, week, diffuse, cylindrical, branched, solid, hairy and green. **L.** Alternate, linear, oblong-lanceolate, obtuse or mucronate. **Fl.** White or light pink. **Fr.** Capsule, scarious, ovate. **S.** 2-4, brownish-black.

79. *Cordia dichotoma* Forst. f.

Family: Boraginaceae

Sleshamantaka

Tree

Plant

Stem

Leaves

Flower

Unripe fruit

Ripe fruit

Dry fruits

Seeds

Key Characteristics for Identification

H. A medium sized tree. **S.** Short crooked, wood soft, light grey. **L.** Glabrous, elliptic-lanceolate to broad ovate. **Fl.** White, pentamerous, in terminal cymes. **Fr.** Drupes, globose, green when unripe and yellowish brown, pink shining with a viscid sweetish transparent pulp. **S.** Hard, rounded.

80. *Cordia rothii* Roem. & Schult.

Laghu Sleshamantaka

Family: Boraginaceae

Tree

Plant

Stem

Branch Full of Fruits

Leaves

Ripe fruit

Flowers

Key Characteristics for Identification

H. A medium sized tree or shrub. **S.** Bark deeply furrowed. **L.** Oblanceolate to oblong, entire to subundulate, coriaceous to scabrid, cuneate subacute. **Fl.** White, in terminal or axillary cymes. **Fr.** Drupes, ovoid, mucronate, yellow to redish-brown, pellucid pulpy, 1-seeded.

81. *Coriandrum sativum* Linn.

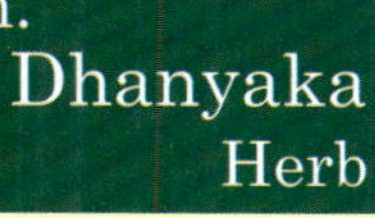

Dhanyaka

Family: Apiaceae

Herb

Plant

Leaves

Leaves

Flowers

Fruits

Seeds

Key Characteristics for Identification

H. A glabrous, aromatic herb. **S.** Angular, branched, solid. **L.** Alternate, compound decompound, pinnae narrow, entire, acute, aromatic. **Fl.** Small, white or pinkish purple, pentamerous. **Fr.** Cremocarp, yellowish brown, globular and 8-10 ribed, containing a seed. **S.** Rounded.

82. *Crataeva magna* Buch.-Ham. Varuna

Family: Capparaceae

Tree

Tree (Inset Tree at Flowering)

Stem

Leaves

Flowers
(Inset Closeup Flower)

Ripe Fruits

Unripe Fruits

Seeds

Dry Stem Bark

Key Characteristics for Identification

H. A medium sized tree. **S.** With brown bark, with longitudinal furrows. **L.** Digitately 3-foliate, leaflets, ovate, lanceolate or obovate, acute or acuminate, glabrous. **Fl.** Large, greenish white or pale yellow, sweet smell. **Fr.** Berry, ovoid, fleshy, woody, red when ripe. **S.** Brown, reniform.

83. *Crinum latifolium* Linn. Sudarsan

Family: Amaryllidaceae Herb

Plant and Leaves

Flower

Flowers

Roots and Bulbs

Key Characteristics for Identification

H. A perennial herb. **B.** Large, subglobose. **S.** Absent. **L.** Numerous, lorate, acuminate, bright green, margins slightly scabrous, stout, tinged with purple. **Fl.** White, fragrant, streaked or tinged with purple, in 10-20-flowered umbles **Fr.** Subglobose. **S.** 12 angled.

84. *Crotalaria juncea* Linn.

Shana

Family: Papilionaceae

Herb

Plant

Leaves

Flowers (Inset Close up Flower)

Fruits

Seeds

Key Characteristics for Identification

H. A hairy herb or undershurb. **S.** Striated, cylindrical. **L.** Linear to oblong-lanceolate, obtuse or sub-acute, apiculate. **Fl.** Large, bright yellow, in erect terminal and lateral 12-20 flowered racemes. **Fr.** Pods, cylindrical, 10-15 seeded. **S.** Reniform, green to grayish, shining.

85. *Cucumis maculata* Roxb. Kachara

Family: Cucurbitaceae Climber

Plant Fruit

Leaves

Flowers

Key Characteristics for Identification

H. A creeping climber. **S.** Angular, scabrous, branches hairy. **L.** 10-12.5 cm long and 15-18.5 cm broad, rough. **Fl.** Yellow, small. **Fr.** 2.5-5.5 cm long, or 10-15 cm long in other variety, colour dark green or pink, with 10 horizontal lines, sour-sweet or much bitter, fragrant. **S.** Small, white, flat.

86. *Cucumis melo* Linn. Kharbuja

Family: Cucurbitaceae

Climber

Plant

Leaves

Flower

Fruits (Inset Fruit variety)

Fruit

Seeds

Key Characteristics for Identification

H. An creeper. **St.** Creeping, angular, scabrous. **L.** Simple, soft hairy, orbicular-reniform, 5 angled or lobed, scabrous on both surfaces, hairs, lobes not deep nor acute. **Fl.** Yellow, unisexual. **Fruit:** Spherical ovoid elongated or contorted, glabrous, with longitudinal lines. **Se.** Flat, white.

87. *Cucumis sativus* Linn. Khira Kakari

Family: Cucurbitaceae

Climber

Plant

Leaves

Flower

Fruits

Key Characteristics for Identification

H. A hairy trailing or climbing herb. **St.** Hairy, angular. **L.** Simple, alternate, deeply cordate, 3-5 lobed, margin denticulate. **Fl.** Yellow, males clustered, females solitary, thickly covered with very bulbous-based hairs. **Fr.** Cylindrical pepo of varying sizes and forms. **Se.** Cream or white, flat.

88. *Curculigo orchioides* Gaertn.

Family: Amaryllidaceae

Kali-Mushali

Herb

Plant

Leaves

Flowers (Inset Closeup Flowers)

Dry Tuber Pieces

Key Characteristics for Identification

H. A tuberous herb. **R.** Short or elongate stout. Outer surface brown-black, inner white. **L.** Basa, crowded on the short stem, membranous, glabrous, linear or linear-lanceolate. 3-5. **Fl.** Bright yellow, in small racemes. **Fr.** Capsule, oblong, hairy. **Se.** Black, shining, oblong or ovoid.

89. *Curcuma longa* Linn.

Family: Zingiberaceae

Haridra

Herb

Plant

Leaves

Flowers (Inset Closeup of Flower)

Fresh Tubers

Dry Tubers

Key Characteristics for Identification

H. A tall herb. **St.** Short and tufts of erect leaves. **R.** Large, ovoid, cylindric tubers, orange-coloured inside, have distinct smell and taste. **L.** Simple, very large, oblong-lanceolate. **Fl.** Spikes 10-15 cm long, flowers pale-yellow. **Se.** Black and shining with large, lacerate and white aril.

90. *Cuscuta reflexa* Roxb. — Amarbel

Family: Convalvulaceae — Climber

Plant

Stem (Inset Stems)

Fruits

Key Characteristics for Identification

H. A leafless climber, intertwining parasite on shrub and trees. The filiform yellowish branches freely swing in the air to and fro. **St.** Herbaceous, long, aerial, weak, cylindrical, solid, glabrous, yellow or greenish-yellow. **L.** Absent. **Fl.** Small, white. **Fruit:** Capsule globose. **S.** Black, glabrous.

91. *Cynodon dactylon* Pers. Durva

Family: Poaceae Herb

Plant

Flowers

Leaves

Key Characteristics for Identification

H. A creeping glabrous herb. **St.** Slender, wiry, rooting at every node forming matted tufts, erect or ascending. **L.** 2-10 cm by 1.25-3 mm, narrowly linear or lanceolate, acute, glacous-green, smooth. **Fl.** Inflorescence terminal spikes, 2-6, green or purplish. **Fr.** Grains, oblong.

92. *Cyperus scariosus* R.Br. Nagarmotha

Family: Cyperaceae

Herb

Plant

Flowers

Leaves

Inflorescence

Tubers

Dry Tubers

Key Characteristics for Identification

H. A glabrous herb. **R.** Tuberous fragrant, outer surface black, inner portion whitish brown, with annular rings. **St.** Erect, triangular, non-branched, solid, smooth, green. **L.** Crowded in the upper part of stem, week, glabrous. **Fl.** Dry, greenish-brown. **Fr.** Achene, obovoid, grayish-black.

93. *Dalbergia sissoo* Roxb. ex DC. Sisama

Family: Papilionaceae

Tree

Tree

Stem

Leaves

Flowers

Fruits

Key Characteristics for Identification

H. A large evergreen tree. **St.** With a crooked trunk light crown and pubescent or tomentose young parts. **L.** Pale green, 3-5 foliate, leaflets alternate, broadly ovate. **Fl.** Pale yellow, in short axillary panicles, shorter than the leaves. **Fr.** Pods, long-stalked, glabrous, strap shaped. **Se.** 1-4.

94. *Datura innoxia* Mill. Dhatura

Family: Solanaceae Herb

Plant

Leaves

Flower

Fruit

Seeds

Key Characteristics for Identification

H. A glabrous herb or shrub. **St.** Herbaceous, green, cylindrical, branched, purplish, glabrous. **L.** Triangular ovate, lanceolate, opposite, one is longer. **Fl.** Infundibuliform, large, white inside, purplish out side. **Fr.** Round, fleshy, pericarp spiny, green. **Se.** Yellowish-brown, flat.

95. *Delonix regia* Rafin. Gulmohar

Family: Ceasalpiniaceae

Tree

Plant

Leaves

Flowers (Inset Closeup of Flower)

Stem

Fruits

Seeds

Key Characteristics for Identification

H. A medium sized tree. **St.** Brown, glabrous, bark smooth, soft, thick. **L.** Compound, abruptly 2-pinnate, pinnae 4-8 pairs, opposite, leaflets 10-20 pairs, linear-oblong, glabrous. **Fl.** Fragrant, yellow-red, in terminal racemes. **Fr.** Pods, 12.5-18 by 2-3.2 cm, glabrous. **Se.** 4-8 in a pod.

96. *Dendrophthoe falcata* (Linn.f.) Etting

Family: Loranthaceae

Banda

Herb

Host Tree (Inset : Plant on Host Tree)

Leaves

Flowers

Fruits

Flowers

Dry Flowers

Key Characteristics for Identification

H. A glabrous, parasite. On the trees of *Madhuka indica.* **St.** Bark grey, smooth, young parts glabrous. **L.** Thick, coriaceous, opposite, ovate, elliptic, or linear-oblong. **Fl.** Orange-red or scarlet in short spreading, stout, axillary racemes. **Fr.** Berries, ovoid or oblong, pink, smooth.

97. *Desmodium gangeticum* D.C.

Family: Papilionaceae

Shalaparni
Herb

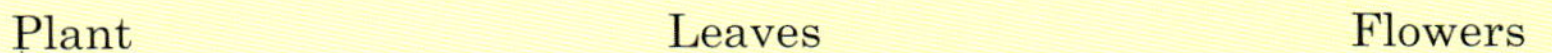

Plant Leaves Flowers

Fruits

Seeds

Key Characteristics for Identification

H. A undershrub. **St.** Pubescent, irregularly angled, short. **L.** Unifoliate, simple, entire, obtuse or acute, ovate to oblong-ovate. **Fl.** Pink, white or purple. **Fr.** Pods, sessile, pubescent, curved, falcate, 6-8 jointed, joints separating when ripe. **Se.** Compressed reniform

98. *Dioscorea bulbifera* Linn.

Family: Dioscoreaceae

Varahikanda
Climber

Plant

Leaves

Fruits (Inset Dry Fruits)

Male Flowers (Closeup of Male Flower)

Raw Tuber Piece

Tubers

Female Flower

Bulbil

Key Characteristics for Identification

H. A twining herb. **Tuber:** Solitary, globose to pyriform, covered with long roots. **St.** Twining to the left, glabrous. **L.** Usually alternate, ovate-cordate, acuminate. **Fl.** Male spikes, clustered, axillary. Stamens 6. Female spikes, in axillary clusters of 2-5. **Fr.** Capsules. **Se.** Winged

99. *Diospyros malabarica* (Desr.) Kostel.

Family: Ebanaceae

Tendu

Tree

Plant

Stem and Leaves

Leaves

Ripe Fruits

Seeds

Key Characteristics for Identification

H. A evergreen tree **St.** Bark thick, smooth, dark grey or black. **L.** Simple, alternate or sub opposite, coriaceous, glabrous, shining, oblong-lanceolate to oblong. **Fl.** Unisexual, yellowish. **Fr.** Ovoid or globose, glabrous, yellow when ripe, with calyx. **Se.** Oblong, reniform, brownish.

100. *Diplocyclos palmatus* (Linn.) Jeffrey

Family: Cucurbitaceae

Shivalingi

Climber

Plant

Leaves

Fruit (Inset: Close up of Fruit)

Flower

Seeds

Ripe fruits

Key Characteristics for Identification

H. A tendril climber. **St.** Grooved, glabrous, tendrils bifid, slender. **L.** 5-lobed, the lobes oblong-lanceolate. **Fl.** Yellow, unisexual, males 3-6, female solitary or few. **Fr.** Berry, globose, smooth, blueish-green when unripe, brick-red when ripe, with white vertical lines. **Se.** Yellowish-brown.

101. *Dolichondrone falcata* Seem.

Family: Bignoniaceae

Meshashringi
Tree

Tree (Inset Tree at Fruiting)

Leaves

Fruit

Flowers (Inset : Closeup of Flower)

Key Characteristics for Identification

H. A small deciduous tree. **St.** Grey. **L.** 1-pinnate, leaflets ovate, suborbicular or obovate, abruptly acuminate, entire. **Fl.** In terminal, few flowered raceme. **Fr.** Capsules, falcate, 2-valved, glabrous. **Se.** Broadly 4-angular, winged at the both ends.

102. *Echinops echinatus* Roxb.

Family: Asteraceae

Untakatela

Herb

Plant

Fruits and Spines

Leaves

Flower

Seeds

Key Characteristics for Identification

H. A much branched rigid herb. **St.** Branches widely spreading from the base, white with cottony pubescent. **L.** Sessile, glabrous or minutely scaberulous above, oblong, deeply pinnatifid, the lobes triangular and oblong. **Fl.** Balls of heads blue or white, spines 2.5 cm long. **Fr.** Achene

103. *Eclipta prostrata* Linn. Bhringaraj

Family: Asteraceae Herb

Plant

Leaves and Flower

Ripe Fruit

Dry Fruits

Fruit

Seeds

Key Characteristics for Identification

H. A much branched herb. **St.** Cylindrical, hairy and light brown. **L.** Opposite, elliptic-lanceolate, crenulate, acute, hairy. **Fl.** Flowers white in heads, heads 6.25-8 mm **Fr.** Cypsella, oblong-turbinate and tuberculate with thickened margins and about 2 mm long. **Se.** Black, small, ovate- long.

104. *Elaeocarpus sphaericus* (Gaertn.) K. Schum.

Rudrakhsa

Family: Elaeocarpaceae

Tree

Tree

Leaves

Stem

Flowering with Fruiting

Flower

Inflorescence

Fruits

Fruit and Seeds

Raw dry fruits

Key Characteristics for Identification

H. A medium-sized evergreen tree. **St.** With a spreading handsome crown. **L.** Oblong-lanceolate, acute or acuminate, glabrous. **Fl.** White, in dense racemes in old leaf axils, opposite each petal. **Fr.** Drupes, globose or obovoid, purple. **Se.** Stone tubercled, longitudinally grooved, 5-seeded.

105. *Erythrina variegata* Linn.

Family: Papilionaceae

Kantaki Palasha

Tree

Tree

Stem

Leaves

Inflorescence

Fruits

Flower

Key Characteristics for Identification

H. A medium sized tree. **St.** Bark yellowish, smooth, shiny, papery, branchlets armed with conical prickles. **L.** Trifoliate, green and glabrous above, rhomboid-ovate, entire or lobed, acute. **Fl.** Large, coral-red. **Fr.** Pods, containing up to 12 seeds. **Se.** Oblong, red to dark purple or brown.

106. *Eucalyptus globules* Labill.

Family: Myrtaceae

Telparni

Tree

Tree

Stem and Leaves

Flowers

Ripe fruit

Unripe fruit

Key Characteristics for Identification

H. A gigantic lofty tree. **St.** Clear straight bole, bark smooth, peeling off in long strips. **L.** Acute or acuminate, opposite or alternate, lanceolate, falcate, curved, rather thick, glaucous. **Fl.** Large, white, axillary, 1-3 together in axils. **Fr.** Hardened capsule, warty, angular. **Se.** Very small.

107. *Euphorbia neriifolia* Auct. non Linn.

Family: Euphorbiaceae

Sehunda

Shrub

Plant (Inset Branch)

Stem (Inset Thorns)

Flowers

Leaves

Fruits

Key Characteristics for Identification

H. A fleshy glabrous, shrub or small tree. **St.** 5-gonous, with pairs of stipular spines on tubercles. Spines sharp, dark brown. **L.** Obovate, oblanceoate-oblong or lanceolate, fleshy, deciduous, shortly acute. **Fl.** Yellowish, 3-7 in a cyme. **Fr.** Tricocus, globose. **Se.** Greenish-brown.

108. *Fagonia indica* Linn. Dhanvayasa

Family: Zygophylaceae

Herb

Plant

Flowers

Leaves

Fruits

Key Characteristics for Identification

H. A spiny herb or undershrub. **St.** Herbaceous, glabrous, solid. **L.** Opposite, 1-3 foliate, striate, slender thorns, leaflets linear or elliptic, on branches always two Leaves, four spines and a flower or fruit. **Fl.** Pale rose, pentamerous. **Fr.** 5 one seeded cocci, glandular, rounded at the base. **Se.** Ovoid.

109. *Ficus arnottiana* Miq. Pahari Pipal

Family: Moraceae

Tree

Tree

Stem

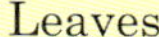

Leaves

Fruits

Key Characteristics for Identification

H. A small tree or shrub. **St.** Bark pale smooth. **L.** Alternate, broadly ovate, base cordate, reddish-brown when dried. **Fl.** Male flowers few, sessile, gall and fertile flowers indistinguishable. **Fr.** Achenes, first white, then red and purple or black at last. **Se.** Small, numerous.

110. *Ficus benghalensis* Linn. Vata

Family: Moraceae Tree

Tree

Arial roots (Inset Stem)

Leaves and fruits

Key Characteristics for Identification

H. A large spreading tree. **St.** Bark greenish-white, young bark is smooth, with longitudinal and transverse rows of lenticles., outer bark easily flakes off. Plenty of latex. **L.** Alternate, leathery, ovate, obtuse, entire, thick, plenty of latex. **Fl.** Globose, brick-red when ripe. **Fr.** Achenes.

111. *Ficus carica* Linn. Anjeer

Family: Moraceae

Shrub

Plant

Stem

Leaves

Lower surface of leaves

Fruits

Dry fruits

Key Characteristics for Identification

H. A shrub or small tree, with latex. **St.** Branchlets pubescent or tomentose. **L.** Cordate, more or less deeply lobed, lobes obtuse. **Fl.** Dioecious, nearly globose receptacles, containing only female flowers. **Fr.** Achenes or drupes, fleshy receptacles that ripen in to purple-brown, pear-shaped fruit.

112. *Ficus hispida* Linn.f. Kakodumber

Family: Moraceae

Shrub

Plant

Stem

Leaves

Fruit

Key Characteristics for Identification

H. A shrub or small tree. **St.** Rough grey bark. **L.** Opposite, membranous, ovate, oblong, or sub-obovate, apiculate or acuminate, surface scabrid. **Fl.** Receptacles paired, pendunculate, globose, scabrous hispid, yellow when ripe. **Fr.** Achenes, hispid, ovoid; style lateral; stigma cylindrical.

113. *Ficus lacor* Buch.-Ham. Plakhsa

Family: Moraceae

Tree

Leaves (Inset: Closeup of Leaves)

Tree

Back surface of leaves

Fruits

Arial roots

Key Characteristics for Identification

H. A large fast growing tree. **St.** Bark grey, smooth, scaly. **L.** Ovate, ovate-oblong, shortly acuminate, base usually rounded, slightly cordate, stipules ovate, acute, pubescent. **Fl.** Male flower few, sessile, gall and fertile flowers. Style of fertile flowers long. **Fr.** Receptacles axillary, in pairs, sessile, globose, whitish.

114. *Ficus racemosa* Linn. Udumber

Family: Moraceae Tree

Plant

Stem

Leaves

Unripe fruit

Ripe fruit

Inner part of unripe fruit

Inner part of ripe fruit

Dry root bark

Close up of ripe fruit

Key Characteristics for Identification

H. An evergreen, large lactiferous tree. **St.** Long, thick, tortuous. **L.** Dark green, ovate or ovate-lanceolate, glabrous. **Fl.** Male, female and gall flowers in one receptacles. **Fr.** Receptacles, sub-globose or pyriform. Figs are smooth, light green when unripe, orange, dull reddish or dark crimson when ripe.

115. *Ficus religiosa* Linn. Pipal

Family: Moraceae Tree

Tree

Stem (Inset Arial Roots)

Leaves

Fruit

Key Characteristics for Identification

H. A large tree. **St.** Bark is grey or ash coloured with thin flakes. Branchlets are drooping. **L.** Bright-green, shining, base broad, cordate, drooping, leathery. **Fl.** The receptacles occurring in pairs, sessile, smooth, green when unripe, purplish when ripe. **Fr.** Achenes, mucilaginous.

116. *Foeniculum vulgare* Mill.

Saunf

Family: Apiaceae

Herb

Plant

Flowers

Seed

Fruits

Key Characteristics for Identification

H. Aromatic, glaucous, stout herb, **Root:** Tap root. **St.** Erect, aerial, hollow, ribbed, herbaceous. **L.** Alternate, compound, 3-4 times pinnetisect, liner or subulate segment. **Fl.** Small, yellow, fragrant. **Fr.** Cremocarp, oblong or ovoid, glabrous, greenish or yellowish brown, with 5 clear unequal strips.

117. *Gloriosa superba* Linn. Kalihari

Family: Liliaceae

Climber

Plant

Leaves

Flower

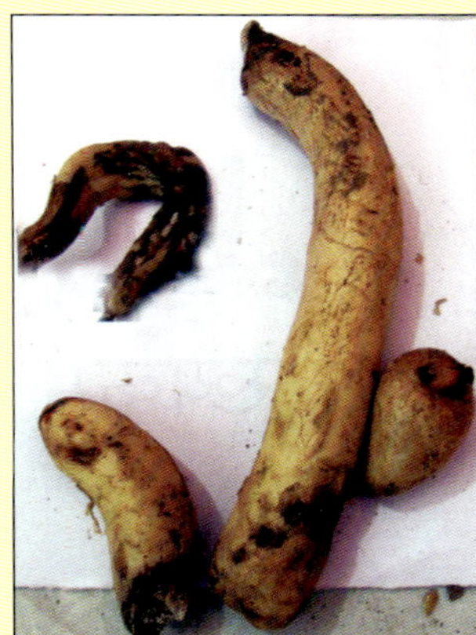

Fresh root

Fruit and seeds

Flowers

Dry piece of roots

Key Characteristics for Identification

H. A handsome climber. **R.** Arched, solid, fleshy-white. **St.** 3-6m long, herbaceous. **L.** Sessile, alternate, ovate-lanceolate. **Fl.** Large, showy, axillary, solitary or sub-corymbose, yellow, passing through orange and scarlet to crimson. **Fr.** Capsule, fusiform or linear-oblong. **Se.** Subglobose, grey, many.

118. *Glycyrrhiza glabra* Linn.

Family: Papilionaceae

Madhuyasti
Herb

Plant

Leaves

Flowers (Inset Closeup of Flowers)

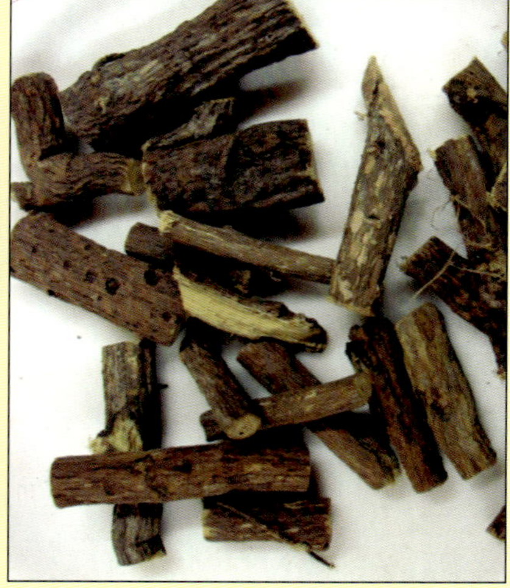

Dry roots

Key Characteristics for Identification

H. A tall herb or undershrub. **Root:** Flexible and fibrous. It has a characteristic pleasant sweet taste. **St.** Pale chocolate brown. **L.** Compound, leaflets 4-7 pairs, oblong to elliptical-lanceolate. **Fl.** Rosy or violet in loose racemes. **Fr.** Pods, oblong to linear, flattened, echinate glandular. **Se.** Many or 2-3, reniform.

119. *Gmelina arborea* Linn. Gambhari

Family: Veranaceae Tree

Tree Stem Leaves

Flower

Ripe fruits and Seed

Unripe fruits

Raw stem bark

Key Characteristics for Identification

H. A moderate sized tree. **St.** Bark whitish grey, exfoliating in thin flakes. **L.** Opposite, broadly ovate, cordate, glandular, acuminate, glabrous, fulvous-tomentose beneath. **Fl.** Orange or brownish-yellow. **Fr.** Drupes, fleshy, ovoid or pyriform, orange-yellow or blackish when ripe. **Se.** 1-3 in a fruit, lenticular.

120. *Gossypium herbaceum* Linn.

Family: Malvaceae

Kapaas

Shrub

Plant

Leaves

Bud

Unripe fruits

Flower

Ripe fruits

Seeds

Key Characteristics for Identification

H. A small shrub. **St.** Rigid and thick, become purple. **L.** Alternate, cleft 3-5-7 lobes, lobes broad, ovate, rotund, suddenly acute or apiculate, leathery. **Fl.** Yellow with purple centre. **Fr.** Capsule, rounded beaked with smooth surface. **Se.** Black, oily, two coats of hairs, white, grey, uniformly over the seed.

121. *Gymnema sylvestre* R.Br. Gurmar

Family: Asclepiadaceae Climber

Plant

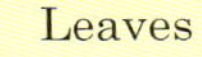

Leaves

Flowers

Fruits

Raw Plant

Seeds

Key Characteristics for Identification

H. A large woody climber. **St.** Young stem and branches pubescent. **L.** Opposite, ovate, elliptic or ovate-lanceolate, or oblong, acute. **Fl.** Yellow, in lateral corymbose cymes. **Fr.** Folicles, linear-lanceolate, attenuated into a beak, glabrous, latex present. **Se.** Ovoid-oblong, flat, brown, with marginal wing.

122. *Helicteres isora* Linn. Marodphali

Family: Sterculaceae — Shrub

Plant

Leaves

Flowers (Inset Closeup of Flower)

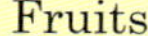

Fruits

Ripe Fruit

Dry Fruits

Seeds

Key Characteristics for Identification

H. A shrub or small tree. **St.** Bark thin, strong and grey. **L.** Simple, alternate, oblong, obovate or roundish. **Fl.** Bright red, bilabiate in axillary clusters of 2-6 together. Pedicels very short, stellately tomentose, subulate, hairy. **Fr.** Folicles 5, beaked, greenish-brown, spirally twisted. **Se.** Numerous, angular.

123. *Hibiscus rosa-sinensis* Linn. Japa

Family: Malvaceae

Shrub

Plant

Leaves

Flowers

Flower variety

Flower variety

White Flower

Key Characteristics for Identification

H. Evergreen, glabrous, showy shrub. **St.** Herbaceous, lower portion woody, glabrous, bark pale-grey or whitish-green. **L.** Bright green, glabrous alternate, ovate, serrate, acute. **Fl.** Red, showy, pentamerous, styles-5, long, passing through the staminal tube stigmatic lobe. **Fr.** Capsule, long, round. **Se.** Numerous.

124. *Holoptelea integrifolia* (Roxb.) Planch.

Family: Ulamaceae

Chirbilva

Tree

Tree (Inset Stem)

Leaves (Inset closeup of Leaf)

Flowers

Unripe Fruit

Ripe Fruit

Seeds

Key Characteristics for Identification

H. A medium sized glabrous tree. **St.** Bark whitish or yellowish grey. **L.** Alternate, simple, elliptic, entire, glabrous,. **Fl.** Greenish-yellow, male and hermaphrodite mixed in short racemes or fascicles near the leaf scars. **Fr.** Samara, sub-orbicular, with membranous reticulately veined wings. **Se.** Shiny, oily, flat.

125. *Hordeum vulgare* Linn.

Family: Poaceae

Jau

Herb

Plant

Fruit

Seeds

Key Characteristics for Identification

H. A tufted, herb. **St.** Green, hollow. **L.** Few, flaccid, linear, acuminate. **Fl.** Spike flattened, 2-ranked, terminal, spikelets sessile, arranged in three on two sides of the rachis. **Fr.** Caryopsis, elliptic, oblong, ovoid, tapering at both ends, dorsally compressed and flattened on the sides with a longitudinal furrow.

126. *Hygrophila auriculata* (K. Schum) Haine

Family: Acanthaceae

Talmakhana
Herb

Plant

Leaves

Seeds

Flowers

Ripe seeds

Spine

Key Characteristics for Identification

H. An erect, hispid herb. **St.** Numerous, sub-quadrangular, grayish-brown, thickened at the nodes, covered with hairs. **L.** Greenish-brown, oblong-lanceolate, hispid on both sides, thorns from the axils of leaves. **Fl.** Purple-blue or rose tint or whitish. **Fr.** Capsules, linear-oblong, pointed. **Se.** Ovate, flat black.

127. *Jasminum gradiflorum* Linn.

Family: Oleaceae

Chameli
Climber

Plant

Leaves

Flower

Key Characteristics for Identification

H. A large glabrous, evergreen twining shrub. **St.** Green, branches ribed. **L.** opposite, imparipinnately compound, ovate-lanceolate, acute or acuminate. **Fl.** Very fragrant, white tinged with pink outside in lax terminal and axillary cymes, calyx lobes long and linear. **Fr.** Berries, elliptic, globose, black when ripe.

128. *Jasminum sambac* (Linn.) Ait.

Family: Oleaceae

Mogra

Herb

Plant

Leaves

Flower

Key Characteristics for Identification

H. A sub-erect or some times climbing shrub. **St.** With pubescent branchlets. **L.** Simple, opposite, ovate or elliptic, acute, entire, nearly glabrous, main nerves-pairs, petioles 3-6 mm long. **Fl.** White, very fragrant, solitary or in 3-flowered terminal cymes. **Fr.** Berries, black when ripe. **Se.** 1-2 in a fruit, black.

129. *Jatropha curcas* Linn. Ratanjot

Family: Euphorbiaceae Shrub

Plant

Leaves

Ripe fruit and Seeds

Ripe fruits

Unripe fruits

Key Characteristics for Identification

H. Large shrub or small tree. **St.** Woody, erect, cylindrical, solid, upper portion red, lower green. **L.** Alternate, broadly ovate, cordate, 3-5 lobed, glabrous, acute, usually palmately. **Fl.** Yellowish green. **Fr.** ovoid, black, splitting into three, 2-valved cocci. **Se.** Dull brownish black, ovoid-oblong, 1.8 cm long.

130. *Kigelia pinnata* (Jacq. DC.)

Gopal Karkati

Family: Bignoniaceae

Tree

Tree

Stem

Leaves

Flower

Fruit

Closeup of Leaves

Dry Fruits

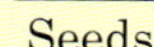

Seeds

Key Characteristics for Identification

H. A much branched, tall tree. **St.** Strait. **L.** 9-50cm long. tomentose. leaflets 5-9, ovate-oblong to elliptic-ovate, coriaceous, entire, acute or mucronate. **Fl.** In lax panicles, pendulous, glabrous, greenish. **Fr.** 30-40 by 10-13cm, woody, gourd like, hanging on cord like penduncles. **Se.** Wingless.

131. *Lagenaria siceraria* (Mol.) Ait.

Family: Cucurbitaceae

Katu Tumbi

Climber

Plant

Fruit

Flower

Fruits (Inset Fruit variety)

Key Characteristics for Identification

H. A large climbing or trailing herb. **St.** Stout, hispid, 5-angled, tendrils long, 2-fid. **L.** Simple, ovate or orbicular, cordate, dentate, 5-angular. **Fl.** White, large, solitary, males long-stalked and the females short-stalked. **Fr.** Large, shape, spherical, bottle or dumb-bell-shaped, rind woody. **Se.** Flat, white.

132. *Lawsonia inermis* Linn.

Family: Lyranthaceae

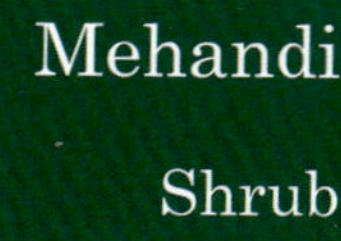

Plant

Leaves

Flowers

Fruits

Dry Fruits

Seeds

Key Characteristics for Identification

H. A glabrous, shrub or small tree. **St.** With 4-gonous lateral branches often ending in spines. **L.** Simple, opposite, entire, lanceolate. **Fl.** White or rose coloured, fragrant. **Fr.** Capsules, globose, tipped with the style and supported by the persistent calyx. **Se.** Brown, pyramidal, externally sub-tuberculate.

133. *Lens culinaris* Medic. Masoor

Family: Pappilionaceae Herb

Plant and Leaves

Fruits

Flower

Seeds

Key Characteristics for Identification

H. A softly pubescent herb. **St.** Branching from the base. **L.** Compound, leaflets 4-6 pairs, opposite, lanceolate, often mucronate. **Fl.** White, rose, red or violate, solitary or in 2-4 flowered racemes. **Fr.** Pods, rhomboid-oblong, compressed. **Se.** Usually two, compressed, varying in colour, grey, pale pinkish.

134. *Leptadenia spartium* Wight.

Family: Asclepiadaceae

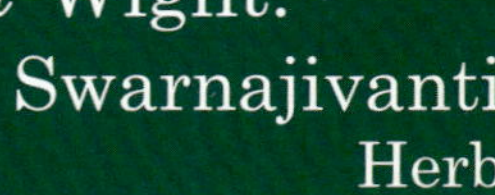

Swarnajivanti

Herb

Stem

Plant

Flower

Key Characteristics for Identification

H. A green herbaceous herb, height 1.25-2m **St.** Green, herbaceous, cylindrical, branches green, thin. **L.** Usually leafless, small in size for little time. **Fl.** Yellow, small.

135. *Limonia acidissima* Linn.

Family: Rutaceae

Kapittha

Tree

Tree

Stem and Leaves

Flowers (Inset Flower)

Fruits

Fruit Pulp

Key Characteristics for Identification

H. A moderate-sized tree. **St.** Short, erect, cylindrical, bearing thorny branches, rough wrinkled bark. **L.** Compound, pinnate, alternate, with small ovate or obovate leaflets, glabrous. **Fl.** Small, light reddish white. **Fr.** Berries, large, globose, rough, grey-white, rind hard, granular. **Se.** Small, flat, in edible pulp.

136. *Limonia crenulata* Roxb.

Family: Rutaceae

Bilvaparni

Tree

Plant

Leaves

Spines

Flowers

Fruits

Key Characteristics for Identification

H. A spinous glabrous shrub or tree. **L.** Pinnate, 2.5-10 cm long, leaflets 2-4 pairs, trapezoid-ovate, obtuse, rarely acute, notched at the tip, crenulate, glabrous. **Fl.** In umbeliform, often leafy raceme, pendunclesa 2-3 together. **Fr.** Berry, 1.2 cm diam., globose, 1-4 seeded, very acid.

137. *Linum usitatissimum* Linn. Alasi

Family: Linaceae Herb

Plant

Leaves

Flower

Seeds

Fruits

Key Characteristics for Identification

H. A glabrous herb. **St.** Solitary or few. **L.** Linear-lanceolate, 3 nerved, acute. **Fl.** Blue or bluish-purple or white, bell-shaped, in terminal corymbose panicles. **Fr.** Capsule, globose, with 5-cells, 2-seeds in each cell, surrounded by persistent sepals. **Se.** Dark brown, ellipsoid, smooth, shiny, oily and mucilaginous.

138. *Luffa acutangula* (Linn.) Roxb.

Family: Cucurbitaceae

Kritvedhan-Turai

Climber

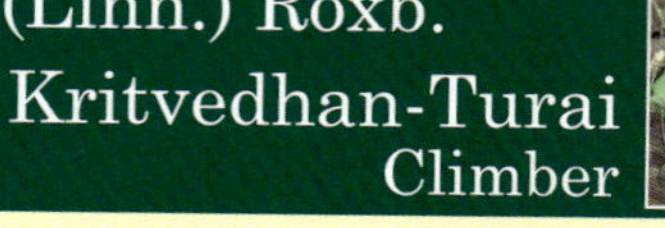

Plant

Flower

Leaves

Fruits

Key Characteristics for Identification

H. A large climber. **St.** 5-angled, glabrous, tendrils usually 3-fid. **L.** Orbicular-cordate, palmately 5-7 lobed or sub-lobate, pale green. **Fl.** Yellow, male in 12-20 flowered axillary racemes, female flowers solitary. **Fr.** Clavate-oblong, very obtuse, smooth, ribbed with 10 sharp angles. **Se.** Black, ovoid-oblong, oily.

139. *Luffa cylendrica* (Linn.) Roem.

Family: Cucurbitaceae

Dhamargava-Gilaki

Climber

Plant

Flower

Leaves

Fruits

Key Characteristics for Identification

H. A large tendrils climber. **St.** Stout, 5-angled, twisted, green. **L.** 5-7 lobed, alternate, acute, hairy, palmately lobed, lobe denticulate. **Fl.** Bright yellow, male flower in clustered 30-45 cm long axillary raceme, female flowers solitary. **Fr.** Pepo, large, cylindrical, smooth, 10- ribbed. **Se.** Black or grey, smooth.

140. *Luffa echinata* Roxb. Devdali

Family: Cucurbitaceae Climber

Plants

Leaves

Fruit

Dry fruit

Flower

Seed

Key Characteristics for Identification

H. A climbing herb. **St.** Slender, branched, furrowed, glabrous, 2-fid. **L.** Reniform-suborbicular, 5-angled or 5-lobed, the lobes rounded or rarely sub-acute at the apex. **Fl.** White. **Fr.** Broadly ellipsoid, clothed with ciliate bristles, green when unripe and brown when ripe, intensely bitter and fibrous. **Se.** Flat, oval.

141. *Madhuca indica* (Koen.) Macb.

Family: Sapotaceae

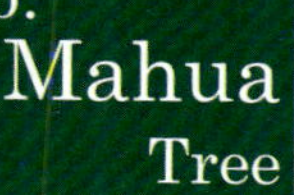

Mahua

Tree

Tree Stem Leaves

Dry Flowers

Seeds

Flowers Fruits (Closeup Fruit with Style)

Key Characteristics for Identification

H. A large evergreen tree. **St.** With dark grey or brownish scaly bark. **L.** Alternate or sub-opposite, thin, often clustered at the end of the branchlets, elliptic or ovate. **Fl.** Pale yellow and fleshy, fleshy, pale yellow, sweetish, edible. **Fr.** Berry, ovoid, yellow when ripe, edible. **Se.** Ovoid, ellipsoid, brown.

142. *Maerua oblongifolia* (Forsk.)

Family: Capparidaceae

Hemkand

Shrub

Plant

Leaves

Ripe fruit

Unripe fruit

Flowers (Inset closeup of flower)

Seeds

Key Characteristics for Identification

H. A climbing shrub up to 4m long. **St.** Branches glabrous to scabrous, divaricate, hairs short. **L.** Elliptic-oblong or lanceolate, entire, obtuse. **Fl.** Usualy in dense, corymbose racemes, greenish-white, pedicellate. **Fr.** Cylindrical, torulose or irregularly many knotted, pale brown. **Se.** Globose, tuberculate.

143. *Majorana hortensis* Linn. Marua

Family: Lamiaceae Herb

Plant

Leaves

Inflorescence

Flowers

Key Characteristics for Identification

H. An erect, perinnial, aromatic herb, 60-90 cm high. **St.** Much branched, more or less hairy. **L.** Stalked, ovate or ovate-lanceolate, 3.5-4.5 cm long, 1,5 - 2.5 cm wide slightly toothed, slightly reddish. **Fl.** Small, white or purplish, in terminal clusters, bracts ovate. **Se.** Minute, oval, dark brown.

144. *Mallolotus philippensis* Muell.-Arg.

Kampillak

Family: Euphorbiaceae

Tree

Tree

Leaves

Inflorescence

Fruits

Fruits

Seeds

Key Characteristics for Identification

H. A small tree. **St.** Bark grey or pale brown, rough. **L.** Alternate, variable, ovate or ovate-lanceolate, glabrous. **Fl.** Small in spikes, the males clustered sessile long terminal spikes, the females in short spikes. **Fr.** Capsules, globose, 3-lobed, bright reddish brown. **Se.** One seed, sub-globose, black and smooth.

145. *Malus pumila* Mill. Seb

Family: Rosaceae Tree

Plant

Leaves

Fruits

Flower

Key Characteristics for Identification

H. A deciduous tree. **St.** Erect, clothed with thin white tomentum. **L.** Ovate, acuminate, clustered on branchlets. **Fl.** White, tinged with red, clustered. **Fr.** Drupe, fleshy, greenish-yellow or red-brown, glabrous, edible, globose, crowned by the persistent calyx-lobes. **Se.** Small, ovate, black-brown.

146. *Mangifera indica* Linn. Aam

Family: Anacardiaceae Tree

Tree

Leaves

Flowers

Fruit (Inset Stem)

Fruits

Seeds

Key Characteristics for Identification

H. A large spreading tree. **St.** Straight, stout clear bole and thick, rough, dark grey bark. **L.** Alternate, linear-oblong, or elliptic-lanceolate, acute, dark green, glabrous, margin wavy. **Fl.** Reddish-white or yellowish-green. **Fr.** Drupes, fleshy, green, orange, yellow or red, obliquely pyriform or sub-ovoid. **Se.** Solitary.

147. *Manilkara hexandra* (Roxb.) Dubard.

Khirani

Family: Sapotaceae

Tree

Tree

Stem

Leaves

Back surface of Leaves

Flowers

Fruits and Seeds

Key Characteristics for Identification

H. A glabrous evergreen tree. **St.** Blackish grey, deeply furrowed. **L.** Simple, alternate, elliptic-obovate or oblong, coriaceous, dark green and polished. **Fl.** White or pale-yellow, small. **Fr.** Berries, ovoid or ellipsoid, milky, one seeded, green when unripe, reddish-yellow when ripe. **Se.** Ovoid, brownish-black.

148. *Marsilea minuta* Linn. Chopatiya

Family: Marsileaceae Herb

Plant

Leaves

Key Characteristics for Identification

H. A creeping herb. **St.** Horizontal, thready. **L.** Quadrifoliate, cuneately-obovate, petioles long, 15-25 cm long, slender, flexible, divided into four leaflets, like *Oxalis corni-culata*. **Fr.** Sporocarps bean-like, born on short or long stalks inserted a short distance above the base of the petioles. **Se.** Small, flattened.

149. *Martynia annua* Linn. Kaknasa

Family: Martyniaceae Herb

Plant

Leaves

Fruits

Ripe fruits

Flowers

Flower

Key Characteristics for Identification

H. A clammy-pubescent sticky, herb. **St.** Erect herbaceous. **L.** Large, opposite, cordate, sinuately lobed. **Fl.** Pink or dark purple, drooping, foxglove shaped. **Fr.** Obliquely ovoid drupaceous with the pericarp, endocarp with 2 sharp curved apical spines and a central channel opening between the spines.

150. *Merremia gangetica* Linn. Cufod.

Family: Convolvulaceae

Mushakarni

Herb

Plant and Leaves

Fruits

Closeup of Leaves

Dry Plant

Flower

Key Characteristics for Identification

H. A perennial, creeping herb. **St.** Numerous, filiform. **L.** Simple, reniform or ovate-cordate, glabrous. **Fl.** Yellow, bell-shaped, axillary, solitary or few (2-3) together on a very short penducle; bracts small, ovate, acute, hairy, pedicels short. **Fr.** Capsule,sub-globose, smooth. **Se.** 2-4, glabrous, light brown.

151. *Mesua ferrea* Linn. Nag Keshar

Family: Clusiaceae

Tree

Tree

Leaves

Unripe Fruits (Closeup Ripe Fruits)

Flowers (Close up of Flower)

Dry Flowers

Seeds

Key Characteristics for Identification

H. A glabrous, evergreen tree. **St.** Reddish-brown, smooth, which peels off in thin flakes. **L.** Opposite, thick, coriaceous, oblong-lanceolate, acute or acuminate, red when young, afterwards shining above. **Fl.** Large, white, very fragrant, axillary or terminal. **Fr.** Ovoid, 2-valved. **Se.** Angular, smooth, dark brown.

152. *Michelia champaca* Linn. Champa

Family: Magnoliaceae

Tree

Tree

Leaves

Branch

Flowers

Fruits

Key Characteristics for Identification

H. It is a tall handsome evergreen tree, **St.** Bark smooth grey or brown, inner part reddish. **L.** Glabrous above, shiny, alternate, ovate, oblong-lanceolate or lanceolate. **Fl.** Yellowish to orange, very fragrant. **Fr.** Etaerio of follicals, ovoid or ellipsoid capsules, dark brown. **Se.** Brown rounded on the back.

153. *Mimosa pudica* Linn. Lajjavanti

Family: Mimosaceae Herb

Plant

Leaves

Fruits (Inset Sleep action at Evening)

Young fruits

Flower

Seeds

Key Characteristics for Identification

H. A prickly under shrub. **St.** Herbaceous, branched, hairy. **L.** Alternate, compound, bipnnate, pinnae 2-4, with 10-20 pairs of leaflets, oblong, acute, hairy, sensitive to touch. **Fl.** Pink In globos heads, prickly, usually in axillary pairs. **Fr.** Lomentum, bristly pods, flat, straw coloured. **Se.** Black, round, shiny, flat.

154. *Mimusops elengi* Linn. Maulshree

Family: Sapotaceae Tree

Tree

Leaves (Inset Closeup of Flower)

Fruits

Unripe fruits

Dry Stem Bark

Flowers

Key Characteristics for Identification

H. A glabrous evergreen Tree. **St.** Short trunk, bark dark grey fissured and dense spreading crown. **L.** Oblong, elliptic or oblaceolate, glabrous. **Fl.** White, fragrant, star shaped, axillary, solitary or in fascicles of 2-6. **Fr.** Berry, ovoid or ellipsoid, yellow when ripe, 1-2 seeded. **Se.** Solitary, ovoid, brown, shiny.

155. *Mitragyna parvifolia* (Roxb.) Korth

Family: Rubiaceae

Kalam

Tree

Tree

Stem

Leaves

Closeup of Leaves

Flowers (Closeup of Flower)

Fruits (Closeup of Fruits)

Key Characteristics for Identification

H. A medium to large sized evergreen tree. **St.** With rounded crown and light grey smooth bark. **L.** Simple, opposite, elliptic, suborbicular, or obovate, glabrous. **Fl.** Greenish-yellow, fragrant, penducles solitary terminal, each carrying a globose head of flowers . **Fr.** Capsules, oblong. **Se.** Globose, hirsuit.

156. *Momordica charanita* Linn.

Family: Cucurbitaceae

Karela

Climber

Plant

Leaves

Flower

Fruits (Inset closeup of Fruits)

Key Characteristics for Identification

H. A monoecious climber. **St.** Long, angled and grooved, hairy, tendrils simple. **L.** Simple, divided in to 5-7 lobes, the lobes acute or subacute, apiculate. **Fl.** Yellow, monoecious. **Fr.** Capsules, bright green when unripe and yellow when ripe, fusiform or oblong-ellipsoid, ribbed and beaked. **Se.** Numerous, dull white.

157. *Moringa oleifera* Lam. Shobhanjana

Family: Moringaceae

Tree

Tree

Stem

Leaves

Fruits

Flowers

Seeds

Gum

Key Characteristics for Identification

H. A small or medium sized tree. **St.** Bark grey corky, wood soft, branches easily breakable. **L.** Tripinnate, rachis slender, thickened, the lateral eliptic, the terminal obovate and slightly longer. **Fl.** White, fragrant, in large puberlous axillary panicles. **Fr.** Pods, pendulous, greenish, triangular, 9-ribbed. **Se.** Trigonous.

158. *Morus alba* Linn. Shahatoot

Family: Moraceae Tree

Tree (Inset Stem)

Leaves (Inset Back surface of Leaves)

Flowers

Fruits (Inset closeup of Fruit)

Key Characteristics for Identification

H. A medium sized tree. **St.** Upper herbaceous, lower woody, bark dark grayish- brown, rough, having vertical fissure. **L.** Alternate, ovate or broadly ovate, glabrous. **Fl.** Greenish, male catkins cylindrical or ovoid; female catkins ovoid. **Fr.** Syncarp of many drupes, white to pinkish-white or dark purple.

159. *Mucana pruriens* (Linn.) DC.

Family: Papilionaceae

Kavaanch

Climber

Plant

Leaves

Flowers

Fruits

Seeds

Key Characteristics for Identification

H. A twining, climber. **St.** Slender, hairy. **L.** Trifoliate, leaflets ovate, elliptic or rhomboid ovate. **Fl.** Purple, 6-30 flowered racemes; the flowers solitary or 2-3. **Fr.** Pods, turgid, ribbed, curved, densely clothed with irritant pale brown or grey, turning steel grey bristles, 4-6 seeded. **Se.** Ovoid, black or dull black or white.

160. *Murraya koenigii* (Linn.) Spreng.

Family: Rutaceae

Meetha Neem

Tree

Plant (Inset Stem)

Leaves

Ripe fruit

Unripe fruit

Flowers

Fruits

Key Characteristics for Identification

H. Small tree. **St.** Dark grey and crowded spreading dark green foliage. **L.** Alternate, compound, unipinnate, ovate, entire, glabrous. cm. long. **Fl.** White, fragrant in panicle cyme, petals-5, gland dotted, stamens-10. **Fr.** Berry, round or ovate, sub-globose or ellipsoid, purplish-black when ripe, 2-seeded.

161. *Musa paradisiaca* Linn.

Family: Musaceae

Kela

Shrub

Plant

Leaves (Inset Stem)

Flowers (Inset Flower)

Ripe Fruits

Unripe Fruits

Key Characteristics for Identification

H. A tall stoloniferous herb. **St.** Aerial. **L.** Oblong, alternate, glabrous, dark green. **Fl.** Brownish red, truncate at base, drooping, females at the bottom and males at the top. **Fr.** Elongated berries, oblong to fusiform, fleshy, triangular, green when unripe and golden yellow on ripening, pulp white and sweet.

162. *Nelumbo nucifera* Gaertn. Kamal

Family: Nymphaeaceae

Herb

Plant and Leaves

Roots (Inset Dry Fruits)

Fruits

Flowers (Inset Inner part of Flower)

Dry Seeds

Seeds

Key Characteristics for Identification

H. A aquatic handsome herb. **St.** Creeping, elongated, branched. **L.** Membranous, orbicular, concave, exactly peltate, entire, radiately nerved, glaucous, glabrous. **Fl.** Solitary, white or rosy, with a yellow obconical spongy torus, becoming enlarged in fruit. **Fr.** Achenes, ovoid, glabrous. **Se.** 8-30, black.

163. *Nerium indicum* Mill.

Family: Apocynaceae

Kaner Shweta

Shrub

Plant and Leaves

Fruits

Flowers

Flowers (Inset variety of Flower)

Key Characteristics for Identification

H. A glabrous evergreen shrub. **St.** Herbaceous, rough, with straight branches, nodes swollen, hyaline latex present. **L.** Dark green, shiny above, three leaves in a whorls, linear-lanceolate, spiny, glabrous, latex present. **Fl.** Red, rosy or white, fragrant. **Fr.** Follicle, cylindrical. **Se.** Numerous, tipped with coma.

164. *Nicotiana tabacum* Linn. Tambaku

Family: Solanaceae

Herb

Plant

Leaves

Flowers

Ripe Fruits

Unripe Fruits

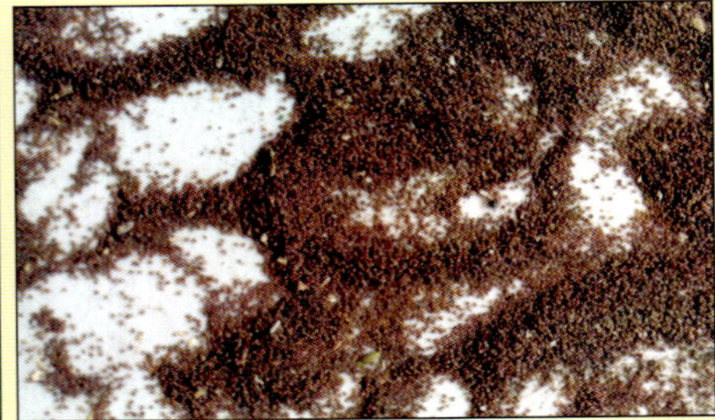

Seeds

Key Characteristics for Identification

H. An viscidly pubescent herb. **St.** Terete, cylindrical. **L.** Simple, large, ovate, oblong-lanceolate or elliptic, acuminate. **Fl.** Rosy, light red or white, 4-5 cm long, pedicelled bracteate, in many flowered panicled racemes. **Fr.** Capsule, 1.5-1.8 cm, narrowly elliptic, ovoid. **Se.** Many, brown, with fluted ridges.

165. *Nyctanthes arbor-tristis* Linn.

Harsingar

Family: Nyctanthaceae

Tree

Plant (Inset Closeup of Stem)

Leaves

Fruits (closeup of fruit)

Flowers

Flower Buds

Flower (Inset Flower side view)

Key Characteristics for Identification

H. A large shrub or small tree. **St.** With grey or greenish-white rough bark, branchlets quandrangular. **L.** Ovate, cordate or oblong, opposite, acute, rough and scabrous above. **Fl.** Small, fragrant, white with bright orange corolla tube. **Fr.** Capsules, sub-orbicular, compressed. **Se.** Flattened, 2, ovate or orbicular.

166. *Nymphaea pubescens* Willd. Utpal

Family: Nymphaeaceae

Herb

Plant and Leaves

Dry Flowers

Flower (Inset Closeup of Flower)

Key Characteristics for Identification

H. A aquatic herb. **L.** Floating, orbicular or elliptic, sharply sinute-toothed or entire, glabrous on both surfaces, with purple beneath. **Fl.** Solitary, floating, pale violate, less commonly light blue or purple, petals linear-oblong or lanceolate. **Fr.** Berry, spongy, globular, many seeded. **Se.** Reddish-grey, roundish.

167. *Ocimum sanctum* Linn. Tulsi

Family: Lamiaceae Herb

Plant

Leaves

Flowers

Fruits

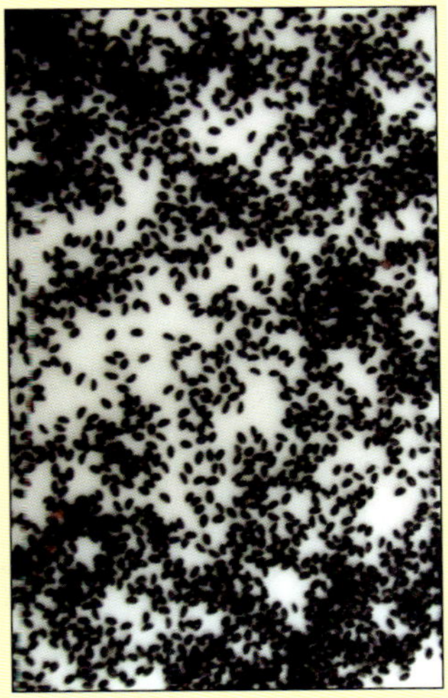

Seeds

Key Characteristics for Identification

H. A much branched herb. **St.** Herbaceous, erect, sub quandrangular, green or purple. **Leaves:** Opposite, ovate, elliptic-oblong, acute or obtuse, pubescent on both sides, aromatic smell present. **Fl.** Purplish or crimson, pentamerous, **Fr.** Nutlets subglobose or broadly ellipsoid. **Se.** Oblong, black, smooth.

168. *Olea europaea* Linn. Jaitoon

Family: Oleaceae Tree

Tree

Leaves (Inset Stem)

Flowers

Fruits

Key Characteristics for Identification

H. An evergreen medium sized tree, height 10 m. **St.** deeply grooved grey trunk. **L.** Small, leathery. **Fl.** Greenish-white, small, clustered. **Fr.** Ovate, green when unripe, turning black when ripe. **Se.** Small, oily, oil tasteless.

169. *Operculina turpethum* (Linn.) Silva Manso

Family: Convolvulaceae

Nishotha

Climber

Plant

Leaves and Flowers (Inset Flower)

Fruits

Dry Fruits with Seed

Roots

Roots variety

Key Characteristics for Identification

H. A large twining herb. **St.** Very long, twining and much twisted. **L.** Ovate or oblong, **Fl.** White tubular, companulate, in few flowered cymes, pendruncles stout, bracts large, lanceolate, stout, pubescent, slightly thickened. **Fr.** Capsules, globose, glabrous or faintly pubescent. **Se.** 4 or less, dull black, glabrous.

170. *Opuntia dillenii* (Ker.-Gawl.) Haw.

Family: Cactaceae

Naagaphani

Shrub

Plant

Flowers

Stem

Fruits and Spines

Key Characteristics for Identification

H. A perennial shrub. **St.** Large spatula-shaped, covered in clusters of spines. **L.** Pale green, conical from a broad base. obovate, dull bluish green, areoles large, bearing 4-6 prickles. **Fl.** Brilliant yellow, tinged with orange. **Fr.** Berry, pyriform, often angular or warty when unripe, deep reddish- purple when ripe.

171. *Orxylum indicum* (Linn.) Vent.

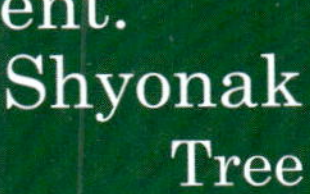

Shyonak

Tree

Family: Bigoniaceae

Plant

Leaves (Inset Stem)

Buds

Fruits (Inset Fruit side view)

Flowers

Seeds of Shyonak

Key Characteristics for Identification

H. A small to medium sized tree. **St.** Bark soft light brown. **L.** Very large, 2-3 pinnate with 5 or more pairs of primary pinnae, 2-4 pairs, ovate or elliptic, glabrous. **Fl.** Lurid to reddish purple outside, dull or pale pinkish yellow within, large, fleshy, shape like trumpet. **Fr.** Capsule, sword shaped. **Se.** Silvery winged.

172. *Oryza sativa* Linn. Chaval

Family: Poaceae Herb

Plant

Leaves

Black Rice

Seeds

Key Characteristics for Identification

H. A erect herb. **St.** Clums hollow, glabrous. **L.** Simple with sheathing bases, long and narrow, linear-lanceolate, striate. **Fl.** Spikelets single, cuneately oblong, pale green, awns 5-13 cm long, golden yellow, reddish-purple, brown, shining. **Fr.** Caryopsis, oblong to ovoid, angular, flattened, red or dirty-white.

173. *Oxalis corniculata* Linn. Changeri

Family: Oxalidaceae

Herb

Plant

Leaves

Flowers

Fruits

Key Characteristics for Identification

H. A procumbent or erect, creeping, acidic, hairy herb. **St.** Diffuse and rooting at nodes, much branched. **L.** Palmately 3-foliate; very slender, pubescent, leaflets obcordate, glabrous. **Fl.** Yellow, 2-5 at each petiole. **Fr.** Capsules, tomentose, oblong or linear-oblong, 5-angled. **Se.** Numerous, broadly ovoid.

174. *Papaver somniferum* Linn. Afeem

Family: Papaveraceae Herb

Plant and Leaves

Fruits

Mature Fruits

Fresh Opium Juice

Opium Juice (after 4 hrs.)

Flowers

Seeds

Key Characteristics for Identification

H. A robust, glaucous green herb. **St.** Scarcely branched. **L.** Simple, sessile, alternate, ovate-oblong, shallowly pinnatifid, lobed, irregularly toothed margin. **Fl.** Large, attractive, scarlet, white, dark stain at the base. **Fr.** Capsule, globose or ovoid, glabrous, stalked. **Se.** White or grayish-brown, reniform.

175. *Pedilium murex* Linn.

Family: Pedaliaceae

Gokharu Bara

Herb

Plant

Leaves

Fruits (Inset Closeup of Fruits)

Dry Fruits

Key Characteristics for Identification

H. A glabrous, succulent herb. **St.** Often slightly rough with scaly glands. **L.** Opposite, simple, pale glacous-green, ovate-oblong, truncate or obtuse, coarsely crenate-serrate or sub-lobate, glabrous above. **Fr.** Pyramidal-ovoid sharp, conical horizontal spines from the angles, 2-chambered. **Se.** Oblong .

176. *Pergularia daemia* (Forsk.) Chiov.

Family: Pariplocaceae

Utaran

Climber

Plant

Leaves

Flower

Fruits

Fruit and Seeds

Key Characteristics for Identification

H. A milky twining herb. **St.** Clothed with spreading hairs. **L.** simple, opposite, ovate or sub-orbicular cordate, acuminate, glabrous, margins ciliate. **Fl.** Greenish-yellow or dull white. **Fr.** Follicles, lanceolate, with long beak and soft spines. **Se.** Ovate, truncate at apex, densely velvety-pubescent on both sides.

177. *Peristrophe bicalyculata* (Retz.) Nees.

Kakjangha

Family: Acanthaceae

Herb

Plant

Leaves

Stem

Flowers (Closeup of Flower)

Key Characteristics for Identification

H. A erect herb. **St.** Herbaceous, erect, usually 6 angular, solid, branched, rough, green, more or less hairy, nodes swollen. **L.** Ovate, acuminate, densely lineolate, more or less hairy above. **Fl.** Small, rosy or purple, pentamerous. **Fr.** Capsule, 4 seeded, purple beaked. **Se.** Orbicular or round, pale-brown or white.

178. *Phaseolus mungo* Linn. non-Roxb. & Auct.

Family: Papilionaceae

Urada

Herb

Plant

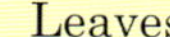

Leaves

Stem

Flowers

Seeds

Key Characteristics for Identification

H. An erect or sub-erect, hairy annual herb, diffusely spreading hairy. **St.** With long twining branches. **L.** Alternate, trifoliate, leaflets elliptic-ovate or oblong-lanceolate, apex acute or acuminate, entire. **Fl.** Yellow, terminal. **Fr.** Pods, cylindrical, very hairy. **Se.** Oblong with square ends, black or greenish-black.

179. *Phaseolus radiatus* Linn. non-Roxb. & Auct.

Munga

Herb

Family: Papilionaceae

Plant

Leaves

Flowers and Fruits

Seeds

Key Characteristics for Identification

H. An sub-erect herbaceous herb. **St.** With twining upper branches and clothed with brownish silky hairs. **L.** Trifoliate, leaflets dark green, entire or rarely trilobed, membranous. **Fl.** Yellow, in clusters of 10-25 on long pedicels. **Fr.** Thin, sub-cylindrical pods, hairy, curved. **Se.** 10-15 green, more or less globular.

180. *Phoenix dactylifera* Linn.

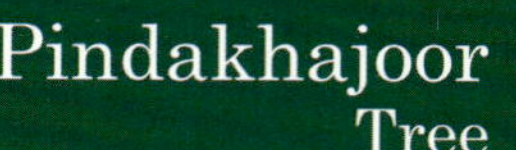

Pindakhajoor

Family: Aracaceae

Tree

Plant

Leaves

Stem and Sister Plants

Ripe Fruits

Unripe Fruits

Dry Fruits

Seeds

Key Characteristics for Identification

H. A tall tree. **St.** The trunk solitary, the base surrounded by a dense mass of root suckers. **L.** Grey, pinnate, up to 5 m long, pinnae 20-40, lower pinnae modified into spines. **Fl.** In long spadices. **Fr.** Berry, oblon, reddish or yellowish brown when ripe, pulp fleshy, sweet. **Se.** Cylindric, hard with a furrow.

181. *Phoenix sylvestris* Linn. (Roxb.)

Family: Aracaceae

Khajoor

Tree

Tree (Inset Stem)

Leaves (Closeup of Leaves)

Ripe Fruit

Unripe Fruit

Key Characteristics for Identification

H. A tall tree. **St.** Woody, erect, cylindrical, Base surrounded by a dense mass of root suckers. **L.** Pinnate, grayish-green, quite glabrous, leaflets modified into spines. **Fl.** Fragrant, unisexual. **Fr.** Berry, oblong-ellipsoid, reddish or yellowish-brown when ripe. **Se.** Round at both ends, cylindrical, hard with a furrow.

182. *Phyla nodiflora* (L.) E. Greene.

Family: Veranaceae

Jal Pippali

Herb

Plant

Leaves

Flower

Key Characteristics for Identification

H. A creeping herb. **St.** Rooting at the nodes and sub-quandrangular stem and branches. **L.** Opposite, cuneate-spathulate, or obovoid. **Fl.** White or pale pink, densely packed in axillary spikes, globose, usually single, broadly elliptic or obovate with a cuneate base, mucronate, glabrous. **Fr.** Globose-oblong, dry.

183. *Phyllanthus emblica* Linn.

Family: Euphorbiaceae

Aanwala
Tree

Plant

Stem

Leaves

Flowers (Inset Closeup of Flower)

Fruits

Dry Fruits

Seeds

Key Characteristics for Identification

H. A small to medium sized tree. **St.** With crooked trunk and thin light grey bark exfoliating. **L.** Light green, glabrous. **Fl.** Greenish-yellow, in axillary fascicles, fragrant. Male numerous, few. **Fr.** On branches, fleshy, globose, with 6 obscure vertical furrows, green and pale yellow, shiny. **Se.** 6, 3-gonous, oily.

184. *Phylanthus fraternus* Webster & Amaru

Family: Euphorbiaceae

Bhumyamalaki

Herb

Plant

Leaves

Flower

Fruits

Seed

Key Characteristics for Identification

H. Erect glabrous herb. **St.** Herbaceous, erect, cylindrical, branched, solid, smooth and green. **L.** Distichous, elliptic-oblong or linear-oblong, alternate, elliptical, glabrous, **Fl.** Yellowish, greenish or whitish, male 1-3, female solitary, **Fr.** Capsule, globose, smooth, **Se.** Pale brown, 3-gonous, 6-7 ribes.

185. *Pisum sativum* Linn.

Matar

Family: Fabaceae

Herb

Plant

Fruit

Flowers

Fruits

Seeds

Key Characteristics for Identification

H. An annual herbaceous tendril climber. **St.** Hollow. **L.** Pinnately compound, leaflets entire, 3-4 by 2-3 cm; stipules auricled. **Fl.** White, red or pink. **Fr.** Pods, straight or curved, 5-10 cm long. **Se.** Globular, smooth, 6-8 mm diam., green, yellowish green, bluish green, white, grey or brown, 6-9 per pod.

186. *Plantago ovata* Forsk. Isabagol

Family: Plantoginaceae Herb

Plant

Leaves

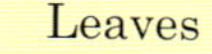

Flowers

Seeds

Key Characteristics for Identification

H. A softly hairy or woolly herb. **St.** none. **L.** Linear or filiform, finely acuminate, entire or distantly toothed, glabrous or pubescent. **Fl.** Small, in ovoid or cylindric spikes, glabrous. **Fr.** Capsule, ellipsoid obtuse, membranous, glabrous. **Se.** Smooth, yellowish brown, ovoid-oblong, boat-shaped.

187. *Plumbago zeylanica* Linn. Chitrak

Family: Plumbaginaceae

Herb

Plant

Leaves

Flowers (Inset Closeup of Flower)

Dry Roots

Fruits

Key Characteristics for Identification

H. A straggling, pretty undershrub. **R.** Light yellowish brown with smooth surface. **St.** Semi-woody, straight, spreading, glabrous. **L.** Alternate, ovate, acute, glabrous, stalk short. **Fl.** White, in terminal and axillary racemes, spikes often branched. **Fr.** Capsule, oblong, pointed, furrowed. **Se.** Small, oblong.

Plant

Leaves (Inset Stem)

Flowers

Inflorance

Ripe Fruits

Unripe Fruits

Key Characteristics for Identification

H. A medium sized glabrous tree. **St.** Short bole and spreading crown, bark grayish-green or brown. **L.** Alternate, imperipinnate, leaflets opposite, ovate or elliptic or acuminate, glabrous. **Fl.** Lilac or pinkish white, fragrant. **Fr.** Pods, thick, smooth, compressed, indehiscent. **Se.** 1-2, white, reniform or elliptical.

188. *Polyalthia longifolia* Thw.

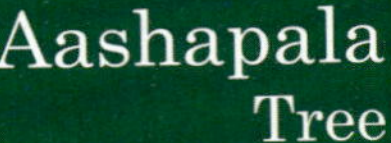

Aashapala

Family: Annonaceae

Tree

Plant (Inset Stem)

Leaves

190. *Psidium guajava* Linn. Amarooda

Family: Myrtaceae

Tree

Tree

Leaves

Stem

Flower

Fruits

Key Characteristics for Identification

H. A Large shrub or small tree. **St.** Bark quite smooth, pale pinkish brown or buff with grey patches, exfoliating in very thin woody flakes. **L.** Opposite, light green, oblong or elliptic-oblong, glabrous. **Fl.** White, fragrant. **Fr.** Berry, fleshy, globose or pyriform, pulp yellowish white or pink. **Se.** White, hard.

187. *Plumbago zeylanica* Linn. Chitrak

Family: Plumbaginaceae

Herb

Plant

Leaves

Flowers (Inset Closeup of Flower)

Dry Roots

Fruits

Key Characteristics for Identification

H. A straggling, pretty undershrub. **R.** Light yellowish brown with smooth surface. **St.** Semi-woody, straight, spreading, glabrous. **L.** Alternate, ovate, acute, glabrous, stalk short. **Fl.** White, in terminal and axillary racemes, spikes often branched. **Fr.** Capsule, oblong, pointed, furrowed. **Se.** Small, oblong.

188. *Polyalthia longifolia* Thw.

Family: Annonaceae

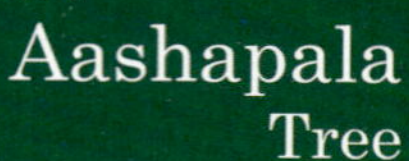

Aashapala

Tree

Plant (Inset Stem)

Leaves

Flowers (Inset Closeup of Flower)

Fruits (Closeup ripe fruit)

Key Characteristics for Identification

H. A tall handsome evergreen tree. **St.** Straight trunk and drooping, glabrous branches, bark smooth, grayish brown. **L.** Lanceolate, tapering to a fine point, shining with undulate margins. **Fl.** Yellowish-green or white, in fascicles or umbels. **Fr.** Carpels, ovoid, one-seeded, purple, glabrous. **Se.** Smooth, shining.

189. *Pongamia pinnata* Pierre. Karanj

Family: Papilionaceae

Tree

Plant

Leaves (Inset Stem)

Flowers

Inflorance

Ripe Fruits

Unripe Fruits

Key Characteristics for Identification

H. A medium sized glabrous tree. **St.** Short bole and spreading crown, bark grayish-green or brown. **L.** Alternate, imperipinnate, leaflets opposite, ovate or elliptic or acuminate, glabrous. **Fl.** Lilac or pinkish white, fragrant. **Fr.** Pods, thick, smooth, compressed, indehiscent. **Se.** 1-2, white, reniform or elliptical.

190. *Psidium guajava* Linn. Amarooda

Family: Myrtaceae Tree

Tree

Leaves

Stem

Flower

Fruits

Key Characteristics for Identification

H. A Large shrub or small tree. **St.** Bark quite smooth, pale pinkish brown or buff with grey patches, exfoliating in very thin woody flakes. **L.** Opposite, light green, oblong or elliptic-oblong, glabrous. **Fl.** White, fragrant. **Fr.** Berry, fleshy, globose or pyriform, pulp yellowish white or pink. **Se.** White, hard.

191. *Pueraria tuberosa* D.C.

Family: Papilionaceae

Vidarikanda

Climber

Plant

Tuber

Leaves

Dry Tuber Pieces

Key Characteristics for Identification

H. A tuberous twining shrub. **T.** Very large, reddish to dark-brown. **St.** Shrubby, woody, porous. **L.** 3-foliate; ovate-oblong, cordate, leaflets sub-coriaceous, terminal large, broadly ovate, oval-rounded, equal sided, lateral ovate-oblong. **Fl.** Blue or purplish-blue. **Fr.** Pods, linear, flat. **Se.** 3-6, reddish-brown.

192. *Punica grantum* Linn. Anaar

Family: Punicaceae Tree

Plant

Leaves (Inset Stem)

Fruits (Inset closeup of Friut)

Flowers

Dry Fruit Bark

Dry Seeds

Seeds

Key Characteristics for Identification

H. A large shrub or small tree. **St.** Dark grey, thin bark and often spinescent branchlets. **L.** Opposite, oblong-lanceolate, oblong-elliptic, glabrous, acute, shining above. **Fl.** Scarlet red, orange, solitary. **Fr.** Globose, crowned by the persistent calyx, rind coriaceous and woody. **Se.** Angular, red, pink or whitish.

193. *Raphanus sativus* Linn. Moolika

Family: Brassicaceae

Herb

Plant and Tuber

Leaves

Flower (Inset Closeup of Flower)

Seeds

Key Characteristics for Identification

H. An annual or biennial herb. **R.** White or brightly coloured, tuberous tap root. **St.** Flexuose, short. **L.** Basal leaves long, latterly pinnate, coarsely toothed, bright green. **Fl.** White or lilac scented in lax racemes. **Fr.** Pods, erect, cylindric, yellow or pale purple, spongy, constricted. **Se.** Ovoid, light brown.

194. *Rauvolfia serpentina* Benth. ex. Kurz.

Sarpagandha

Family: Apocynaceae

Herb

Plant

Fruit

Flowers

Fruits

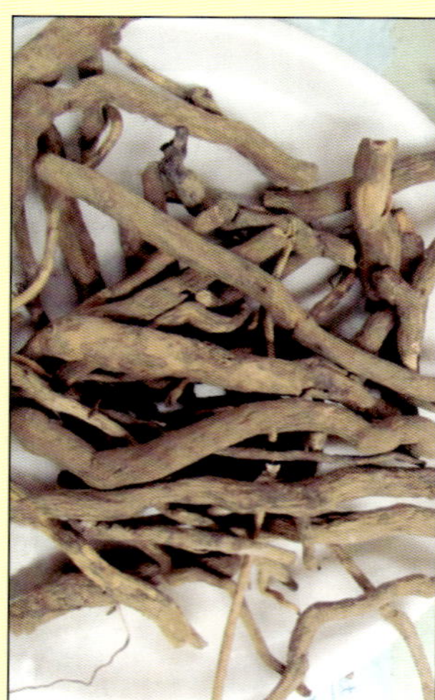

Dry Roots

Key Characteristics for Identification

H. A small shrub or undershrub. **R.** Curved and twisted, brownish, tuberous. **St.** Bark pale brown, corky, fissured. **L.** 3 in a whorl, thin, lanceolate, acute or acuminate, glabrous, bright green. **Fl.** White, often tinged with violate. **Fr.** Drupes of 2 pyrenes, single, purplish-black when ripe. **Se.** Pyrenes rugose, hard.

195. *Rhus parviflora* Roxb. Tintideek

Family: Anacardiaceae

Shrub

Plant Stem

Leaves Fruits

Key Characteristics for Identification

H. An evergreen shrub. **St.** Bark dark brown roughish, exfoliating in small woody scales. **L.** Alternate, trifoliate, tomentose, terminal largest, obovate. **Fl.** Yellowish-green, in terminal panicles the lower branches axillary, pediceles minute. **Fr.** Drupes, 5 mm, ovoid, brown, glabrous, shining. **Se.** Pendulous.

196. *Ricinus communis* Linn. Erand

Family: Euphorbiaceae

Shrub

Plant

Fruit

Fruits

Flowers (Red Variety)

Flowers (White Variety)

Dry Fruits

Dry root pieces

Seeds

Key Characteristics for Identification

H. A tall glabrous, shrub or small tree. **St.** Herbaceous, cylindrical, branched, glabrous, green-purple. **L.** Palmetly lobed, alternate, 5-11 lobed. **Fl.** Monoecious, large, male in upper portion of inflorescence, the females below. **Fr.** A prickly capsule of three 2-valved cocci, 3 seeded, **Se.** Oblong, hard, spotted.

197. *Rosa centifolia* Linn.

Gulab

Family: Rosaceae

Herb

Plant

Stem

Leaves

Fruit

Flower Bud

Flower

Key Characteristics for Identification

H. A small prickly shrub. **St.** With large, hooked prickles and many bristles. **L.** Compound, alternate, leaflets usually five. **Fl.** Usually pink, very fragrant, double on long slender pedicles, terminal, solitary, petals numerous, stamens many, sepal persistent. **Fr.** A fleshy hip enclosing bony achenes. **Se.** Small.

198. *Saccharum officinarum* Linn.

Family: Poaceae

Ikhsu

Shrub

Plant

Stem

Leaves

Dry Leaves (Inset Dry roots)

Flowers

Key Characteristics for Identification

H. A tall herb. **St.** Varying in thickness and colours, with many distinct nodes and internodes, cylindrical, solid juicy, glabrous, coated with wax below the nodes and with sweet juice. **L.** Leaf sheaths smooth, glabrous, green. **Fl.** Panicles pyramidal, long, dense silvery, silky hairs. **Se.** Oblong or sub-globose.

199. *Salix tetrasperma* Roxb. Jal Vetas

Family: Salicaceae

Tree

Tree

Stem

Inflorance

Leaves

Flowers

Key Characteristics for Identification

H. A medium sized tree. **St.** Brown or blackish, with longitudinal furrows, rough. **L.** Lanceolate or ovate-lanceolate, caudate-acuminate, green and glabrous. **Fl.** Male catkins sweet scented, sessile or on leafy shoots, rhachis hairy; Female catkins on leafy shoots. **Fr.** Capsules, glabrous. **Se.** 4-6, brown-blackish.

200. *Salvadora persica* (Burm.f.) Wight & Arn.

Family: Salvadoraceae

Peelu

Tree

Plant

Stem

Leaves

Fruits

Flowers

Key Characteristics for Identification

H. A large shrub or small tree. **St.** Hollow, crooked, short, with soft whitish yellow wood, bark of old stem rugose; branches numerous, dropping, glabrous. **L.** Fleshy,opposite, elliptic-lanceolate or ovate, characteristic smell. **Fl.** Greenish-yellow. **Fr.** Drupes, globose, smooth, red, mucilaginous. **Se.** Small, one.

201. *Sanseveiria roxburghiana* J.&J. Schultes

Nagdaman

Family: Liliaceae

Herb

Plant

Leaves

Ripe Fruit

Unripe Fruit

Key Characteristics for Identification

H. A perennial herb. **St.** Absent. **R.** Creeping. **L.** 6-24 in number, radical, smooth above, rough beneath, strap-shaped or lanceolate, blackish-green, concave, transversely marked with darker green regular bars on both sides. **Fl.** White or light green, 4 in a cluster. **Fr.** Membranous, indehiscent. **Se.** 1-3, globose.

202. *Santalum album* Linn.

Family: Santalaceae

Chandan Swet

Tree

Tree

Stem

Inflorance

Leaves

Flowers

Fruits

Wood Pieces

Key Characteristics for Identification

H. A glabrous tree. **St.** Bark dark grey or brownish-black, rough, with slender drooping branches, the heartwood yellowish-brown, strongly scented. **L.** Opposite, elliptic-lanceolate, sub-acute, thin, glabrous. **Fl.** Brownish-purple, reddish-purple or violate. **Fr.** Drupes, globose, purple-black. **Se.** Hard, globose.

203. *Sapindus laurifolius* Gaertn.

Aritha

Family: Sapindaceae

Tree

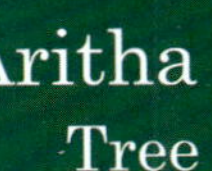

Tree

Stem

Flowers

Leaves

Fruits

Small Fruits

Dry Fruits

Key Characteristics for Identification

H. A small or medium sized deciduous tree. **St.** Bark dark or pale grey, rough, exfoliating in irregular woody scales. **L.** Leaflets 5-8 pairs, opposite or alternate, lanceolate, acuminate, glabrous. Petiolules 2.5-5 mm long. **Fl.** White or purple, numerous. **Fr.** Drupes, smooth, pale brown, globose. **Se.** Globose, black.

204. *Saraca asoca* (Roxb.) De. Wilde.

Family: Ceasalpiniaceae

Ashoka
Tree

Tree

Stem

Leaves

Young Leaves

Flowers (Closeup of Flower)

Key Characteristics for Identification

H. A Medium sized tree. **St.** Erect, bark dark-brown to grey or black. **L.** Pinnate, leaflets 4-6 pairs, oblong-lanceolate, obtuse or acute, quite glabrous, wrinkled. **Fl.** Orange or orange-yellow, fragrant, numerous. **Fr.** Pods, black, linear-oblong, compressed, glabrous, woody, veined. **Se.** 4-8 per pods, ellipsoid-oblong.

205. *Schrebera swietenioides* Roxb.

Family: Oleaceae

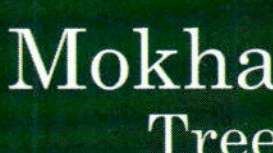

Mokha

Tree

Tree (Inset Stem)

Leaves

Fruit

Open Fruit

Key Characteristics for Identification

H. A moderate sized tree. **St.** With thick grey bark. **L.** Imparipinnate, leaflets opposite, 3-4 pairs, ovate, entire, unequal-sided, thickened at the insertion of leaflets. **Fl.** Yellowish-brown or whitish, fragrant at night, in terminal compound cymes. **Fr.** Capsules, pendulous, pear-shaped, 8-seed. **Se.** Angular, winged.

206. *Sesamum orientale* Linn. Til

Family: Pedaliaceae Herb

Plant

Leaves

Flower

Fruits

Seeds

Key Characteristics for Identification

H. A foetid-glandular herb. **St.** Branching from the base, quadrangular, pubescent. **L.** Simple, large, thin, the lower ones deeply 3-lobed; lobes lanceolate, hairy, upper entire, linear. **Fl.** White or pink or purplish. **Fr.** Capsules, oblong, bluntly 4-gonous, 4-grooved. **Se.** Obovoid, compressed, white, black and red.

207. *Sesbania grandiflora* (L.) Poir.

Family: Papilionaceae

Agatsya

Tree

Plant

Leaves

Flowers (Inset Flower)

Fruits

Key Characteristics for Identification

H. A slender, small tree. **St.** Soft wooded, with spreading branches. **L.** Abruptly pinnate; leaflets 16-30 pairs, linear-oblong or oblong-elliptic, glabrous above. **Fl.** Showy, fleshy, white, pink or crimson. **Fr.** Pods, pendulous, flat, 4-cornered, non-torulose, septate with swollen margins. **Se.** 15-50 per pod, pale.

208. *Sesbania bispinosa* (Linn.) Merr.

Family: Papilionaceae

Jayanti

Shrub

Plant

Stem

Leaves

Fruit

Bud

Seed

Key Characteristics for Identification

H. An annual herb. **St.** Cylindrical, branched, solid, glabrous and green. **L.** Alternate, compound, unipinnate and paripinnate, oblong, entire, glabrous, leaflets 8-20 pairs special smell in leaves. **Fl.** Small, group of 8-10 flowers, yellow or red. **Fr.** Legume, beaked, bent, 20-30 seeded. **Se.** Oblong, shiny, hard.

209. *Sida cordifolia* Linn.

Family: Malvaceae

Balaa

Herb

Plant and Leaves

Flower

Fruits

Dry Fruits

Dry Roots

Seed

Key Characteristics for Identification

H. Herb, hairy, grayish-green, woody. **St.** Herbaceous, aerial, erect, cylindrical. **L.** Alternate, stipulate, orbicular, ovate-oblong or cordate, acute, pubescent, hairy. **Fl.** Yellow or white, bracteate, pedicilate, complete, pentamerous. **Fr.** Schizocarpic, depressed, globose. **Se.** Brown or black, flattened, reniform.

210. *Solanum ferox* Linn. Bari Kantakari

Family: Solanaceae Herb

Plant

Leaves

Ripe Fruit

Flowers (Closeup of Flower)

White Flowers

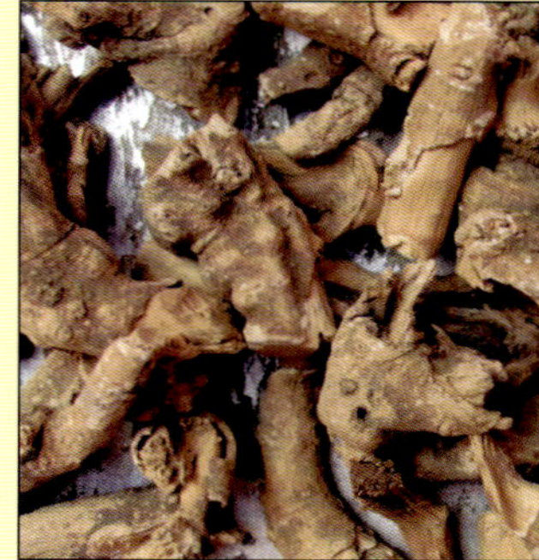

Dry Roots

Unripe Fruit

Seed

Key Characteristics for Identification

H. An armed herb or undershrub. **St.** Herbaceous, armed with straight prickles. **L.** 2 at a node and unequal, elliptic, sinutely or pinnately, armed with long, erect, yellow prickles, unequal-sided. **Fl.** Blue or purple, solitary. **Fr.** Berry, globose, green with white lines unripe and yellow when ripe. **Se.** Glabrous.

211. *Solanum melongena* Linn. Begun

Family: Solanaceae Herb

Plant

Leaves

Flowers (Inset Seeds)

Fruits

Key Characteristics for Identification

H. An erect or suffrutescent, herbaceous, armed or unarmed herb. **St.** Young parts closely tomentose. **L.** Simple, large, entire, lobed, large elliptic. **Fl.** Blue or purple, in clusters of 2-5. **Fr.** Berry, large, glabrous, dark purple, white, yellow, varying in shape and size, capped with persistent calyx. **Se.** Yellow or cream.

212. *Solanum nigrum* Linn. Makoy

Family: Solanaceae Herb

Plant

Leaves

Flowers

Ripe Fruits

Unripe Fruits

Seed

Key Characteristics for Identification

H. An unarmed erect, herb. **St.** Herbaceous, aerial, cylindrical, solid, smooth or puberulous, green with terete or angular. **L.** Dark green, thin, ovate, oblong or ovate-lanceolate, entire. **Fl.** White, pentamerous. **Fr.** Berry, green when unripe and purplish-black or reddish when ripe, **Se.** Smooth, ovoid, yellow.

213. *Solanum surattense* Burm.f.

Choti Kantakari

Herb

Family: Solanaceae

Plant | Leaves

Flowers | Unripe Fruit

Ripe Fruits | Dry Roots | Seed

Key Characteristics for Identification

H. A very prickly diffuse bright green herb. **St.** Woody at base, with zigzag branches that spread close to the ground cover over with strong, sharp, glabrous, yellowish white prickles. **L.** Ovate or ovate-oblong, elliptic. **Fl.** Purple or bluish-purple. **Fr.** Berry, glabrous, globular, yellow or white. **Se.** Glabrous, reniform.

214. *Soymida febrifuga* A. Juss.

Family: Meliaceae

Mansrohini

Tree

Tree

Stem

Stem Bark

Leaves

Fruits

Key Characteristics for Identification

H. A tall tree. **St.** Bark bluish-grey or blackish. **L.** At ends of the branches; leaflets 3-6 pairs, opposite, elliptic or oblong, obtuse, glabrous. **Fl.** Small, greenish-white, in large terminal or axillary divercately branched panicles. **Fr.** Capsules, hard, brown-red, obovoid, 5-celled, 5-valved. **Se.** Numerous, winged.

215. *Sphaeranthus indicus* Linn.

Family: Asteraceae

Gorakhmundi
Herb

Plant

Leaves

Fruits

Dry Fruits

Key Characteristics for Identification

H. An aromatic, prostrate herb. **St.** Stem and branches are cylindrical, glandular, hairy, toothed winged. **L.** Sessile, decurrent, obovate-oblong, glandular-hairy, spinous-serrate or dentate, at the base. **Fl.** Purple or pink, in terminal, globose or ovoid, heads, on solitary and grandular peduncles, **Fr.** Achenes, glabrous.

216. *Sterculia urens* Roxb. Kadaya

Family: Sterculiaceae Tree

Tree

Stem

Leaves (Inset Flowers)

Fruits

Key Characteristics for Identification

H. A deciduous tree. **St.** Strait, bark white, smooth, outer papery peeling off, inner fibrous. **L.** Simple, alternate, entire or palmately lobed or digitately compound, 3-5 lobed. **Fl.** Small, yellow or olive-yellow, numerous. **Fr.** Etaerio of stellately spreading, woody follicles. **Se.** Oblong, dark brown or black, glossy.

217. *Sterospermum personatum*

syn. *chelenoides* (Hassk.) D. Chatterjee

Family: Bignoniaceae   Ghanta Padhal   Tree

Tree (Inset Stem)

Flowers (Inset Flower)

Leaves

Fruits

Seeds

Key Characteristics for Identification

H. A large tree, with many spreading branches. **St.** Bark yellowish, thick. **L.** Imperipinnate, leaflets 3-5 pairs, elliptic, caudate-acuminate, glabrous, often unequal-sided. **Fl.** Fragrant, yellow, in lax dropping terminal panicles. **Fr.** Capsules, subquadrangular, curved, pointed, twisted, brown. **Se.** 2.5 to 3.2cm long.

218. *Sterblus asper* Lour. Shakhotak

Family: Moraceae

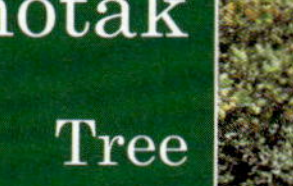

Tree

Plant

Leaves

Closeup of Leaves

Fruits

Key Characteristics for Identification

H. A small rigid tree. **St.** With light grey or greenish rough bark, milky white latex. **L.** Simple, alternate, rhomboid-elliptic, obovate or elliptic-oblong, acute or acuminate. **Fl.** Male flowers in globose heads. Female flowers solitary. **Fr.** Berry, one seeded, yellow when ripe. **Se.** Smooth, round, greenish white.

219. *Syzygium cuminii* (Linn.) Sklles.

Family: Myrtaceae

Jamun

Tree

Tree (Inset Stem)

Leaves

Flowers

Unripe Fruits (Inset Ripe Fruit)

Unripe Fruits

Seeds

Key Characteristics for Identification

H. Tree of 15-30 m height. **St.** Bark light grey, smooth, cylindrical. **L.** Opposite, elliptic-ovate, entire, acute, glabrous. **Fl.** Greenish-white, clustered, fragrant, complete. **Fr.** Drupe, oblong or ovoid-oblong, green when unripe, dark purple when ripe, with pinkish juicy pulp, one seeded. **Se.** Hard, 1-2 cm long.

220. *Tamarindus indica* Linn. Imali

Family: Ceasalpiniaceae Tree

Tree

Stem

Leaves and Flowers

Fruits

Fruits and Seeds

Key Characteristics for Identification

H. Large, evergreen tree. **St.** Bark dark grey having longitudinal fissure and deep cracks. **L.** Up to 15 cm long, leaflets 10-20 pairs, alternate, compound, glabrous. **Fl.** Yellow, striped red. **Fr.** Legume, brownish ash coloured, unripe pods are sour, ripe pods are sour, sweet. 3-12 seeded. **Se.** Hard, brownish-black.

221. *Tecomella undulata* (G.Don.) Seem.

Family: Bigoniaceae

Rohitak

Tree

Tree

Stem

Leaves

Flowers (Closeup of Flower)

Dry Stem Bark

Key Characteristics for Identification

H. A small tree. **St.** With drooping branches and stellately grey-tomentose innovations, otherwise glabrous. **L.** Simple oblong or oblong-lanceolate, obtuse. **Fl.** Orange-yellow, large, showy, **Fr.** Capsules, 20 by 1 cm, curved, linear-oblong, acute, smooth; valves thin. **Se.** 2.5 by 1 cm, discoid and winged.

222. *Tectona grandis* Linn.f. Sagwaan

Family: Verbanaceae Tree

Plant

Leaves

Fruits (Inset Stem)

Flowers

Dry Fruits

Seeds

Key Characteristics for Identification

H. A very large tree. **St.** Bark light brown or grey, having furrows, fluted. **L.** Simple. Opposite, elliptic or obovate, acute or acuminate. **Fl.** White, numerous, small, sweet scented, in terminal cymose panicles. **Fr.** Drupes, hard, bony, globose, enveloped by light brown calyx, somewhat 4-lobed. **Se.** Ovate, white.

223. *Tephrosia purpurea* (L. Pers.)

Family: Papilionaceae

Sharpunkha
Herb

Plant

Leaves

Fruits

Flowers (Inset Closeup Flower)

Ripe Fruits

Seeds

Key Characteristics for Identification

H. A herb or undershrub. **St.** Spreading branched, glabrous. **L.** Linear-subulate, nerved, erect, leaflets 11-21, narrow, oblanceolate, obtuse or retuse, mucronate, glabrous above, fine silky hairs beneath. Leaf breaks lanceolate shape. **Fl.** Bright rosy-purple or violate. **Fr.** Pods, curved, linear. **Se.** Grey and smooth.

224. *Terminalia arjuna* (Roxb.) W.&A.

Family: Combrataceae

Arjun

Tree

Tree

Stem

Leaves

Flowers

Fruits

Ripe Fruits

Dry Stem Bark

Fresh Stem Bark

Key Characteristics for Identification

H. A large tree. **St.** Bark smooth, whitish outside and flesh coloured inside, flaking off in large flat pieces. **L.** Simple, sub-opposite, oblong or elliptic-oblong, obtuse or subacute, coriaceous, pale dull green 10-15 pairs. **Fl.** White, sessile, in terminal panicles. **Fr.** Drupes, ovoid or obovoid-oblong, glabrous, dark-brown.

225. *Terminalia bellirica* Roxb. Bahera

Family: Combrataceae

Tree

Tree (Inset Stem)

Leaves

Flowers

Fruits

Dry Fruits

Key Characteristics for Identification

H. A large buttressed tree. **St.** Brownish grey bark having fissures. **L.** Simple, alternate, coriaceous, broadly elliptic or elliptic-obovate, rounded, both surfaces puberulous when young, glabrous and reticulate when old. **Fl.** Pale greenish yellow. **Fr.** Drupes, ovoid, grey, velvety. **Se.** Thick walled, hard.

226. *Terminalia chebula* Retz. Harad

Family: Combrataceae Tree

Tree (Inset Stem)

Leaves

Flowers

Fruits Julfa

Fruits Kabuli

Dry Fruits

Young Fruits

Ripe Fruits

Unripe Fruits

Key Characteristics for Identification

H. A large tree. **St.** Bark dark brown, young leaves are covered with soft, shining, rust or silvery hairs. **L.** Alternate or sub-opposite, glabrous, ovate or elliptic-oblong, rounded or cordate at base. **Fl.** Yellowish-white. **Fr.** Drupes, obovoid or ovoid, glabrous, shining, yellow to orange-brown. **Se.** Hard, pale yellow.

227. *Thespesia populea* Soland. ex Correa.

Family: Malvaceae

Paras Pipal

Tree

Plant (Inset Stem)

Leaves

Flower

Ripe Fruits

Unripe Fruits

Key Characteristics for Identification

H. A medium to large sized, tree. **St.** With greyish brown, fissured bark. **L.** Ovate, cordate, acuminate, entire, smooth. **Fl.** Yellow with purple base, slowly changing to purple, bell-shaped. **Fr.** Capsules, globose or oblong, brown, 4-seeded, with yellow latex, surrounded at the base by the persistent calyx. **Se.** Ovoid.

228. *Thevetia neriifolia* Juss. Kaner

Family: Apocynaceae Shrub

Plant (Inset Stem)

Leaves

Fruits

Flowers

Key Characteristics for Identification

H. Glabrous shrub, with milky latex. **St.** Herbaceous, smooth and green. **L.** Alternate, linear-lanceolate, acute, glabrous, shining above, narrowed to both ends, milky latex present. **Fl.** Yellow, pentamerous, fragrant. **Fr.** Drupe, obovate, green when unripe, brown when ripe. **Se.** 4, Light brown, specific triangular.

229. *Thuja orientalis* Linn. Mayurpankhi

Family: Cupressaceae

Shrub

Plant (Inset Stem)

Leaves

Fruits (Closeup of Fruit)

Ripe Fruits

Key Characteristics for Identification

H. A evergreen, large tree, but often branched shrub. **St.** Branchlets 2-ranked. **L.** Scalelike, less than 3 mm long, glandular, grooved. **Fl.** Male cones at the terminal of branches of the last year, with 10 scales. erect, the 3-9 pairs of scales fleshy when young, becoming woody strongly hooked at apex. **Se.** Wingless.

230. *Tinospora cordifolia* (Willd.) Miers. ex Hook. & Thoms

Family: Menispermaceae

Giloy

Climber

Plant   Leaves   Fruits (Closeup of Fruits)

Dry Branches   Seeds

Flowers   Branch (Inset Unripe Fruits)

Key Characteristics for Identification

H. A large, glabrous, climber. **St.** Succulent, corky, grooved, warty tubercled, pale, shining or glaucous. **L.** Alternate, broadly ovate, roundish or subdeltoid-cordate. **Fl.** Yellow or greenish-yellow, male clustered, female solitary. **Fr.** Drupes, ovoid, glossy, green when unripe and red when ripe. **Se.** White, warty.

231. *Trachyspermum ammi* (Linn.) Sprague.

Family: Apiaceae

Ajawayan

Herb

Plant

Leaves

Flowers

Unripe Fruits

Seeds

Key Characteristics for Identification

H. A erect herb. **St.** Much branched, slender, striate, glabrous. **L.** Rather distant, 2-3 pinnate, slender. Linear or filiform. **Fl.** Pure white, Small, polygamous, in terminal or lateral peduncles compound umbles. **Fr.** Cremocarps, ovoid, muricate, sub-hispid, aromatic, greyish-brown, ridges distint, small. **Se.** One seed.

232. *Trapa natans* Linn.

Family: Trapaceae

Singhara

Herb

Plant (Inset Flowers)

Leaves (Inset Back of Leaves)

Roots

Fruits

Key Characteristics for Identification

H. An attractive aquatic, herb. **St.** Flexuose, ascending, and green roots. **L.** Alternate, rhomboid, apex triangular. **Fl.** White, few, axillary, solitary, tubular. **Fr.** Nut, obovoid, bony, 4-angular, with a short conical beak, sharp spinous horn on side, green, dark maroon-blackish, one seeded. **Se.** White, starchy.

233. *Tribulus terrestris* Linn.

Family: Zygophylaceae

Gokharu Chota

Herb

Plant

Leaves

Flowers

Fruits

Dry Fruits

Key Characteristics for Identification

H. A prostrate herb. **R.** Grey, fibrous, cylindrical, tough. **St.** Stem and branches pilose. **L.** Opposite, one of each pair smaller than other, hairy; leaflets 3-6 pairs, oblong. **Fl.** Bright yellow, solitary. **Fr.** Schizocarp, globose, 5-angled, coccus having two spines towards the distal half and two nearer the base. **Se.** One.

234. *Trichodesma indicum* R.Br.

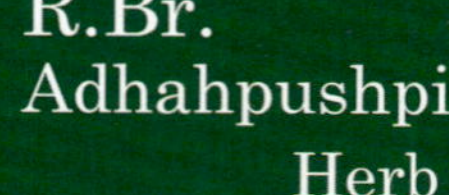

Adhahpushpi

Family: Boraginaceae

Herb

Plant

Leaves

Flowers

Buds

Key Characteristics for Identification

H. A hispid herb. **St.** Erect, herbaceous. **L.** Opposite, ovate or oblong, or lanceolate-oblong, obtuse or sub-acute, clothed above with stiff hairs. **Fl.** Pale violate-blue, changing pink or white on drooping axillary stalks, clothed with white hairs. **Fr.** Nutlets, pyramidal, 4-ribbed, ovoid, smooth, bluish white.

235. *Trichosanthes bracteata* (Lam.) Voigt.

Indrayan Bari

Family: Cucurbitaceae

Climber

Plant

Leaves

Fruits (Closeup of Fruit)

Dry Fruits

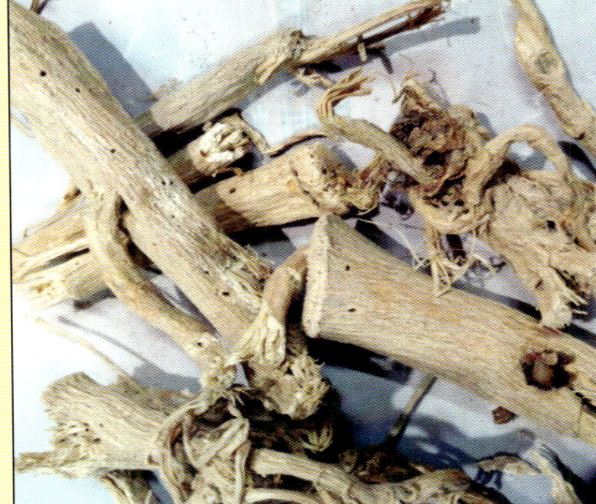

Dry Roots

Key Characteristics for Identification

H. A tendril climber. **St.** Robust, grooved, older light grey, younger smooth, green, tendril 2. **L.** Palmately 3-5 lobed, dark green above,lobes usually ovate-oblong. **Fl.** White, male in axillary racemes. Female solitary. **Fr.** Berry, fleshy, globose, red when ripe, streaked with 10 orange streaks. **Se.** Ellipsoid, smooth.

236. *Trigonella foenum-graecum* Linn.

Family: Papilionaceae

Methi

Herb

Plant (Inset Flower)

Leaves

Fruits

Seeds

Key Characteristics for Identification

H. An aromatic, hairy, herb. **St.** Herbaceous, green, branched, glabrous. **L.** Cauline and ramal, alternate, compound, tripinnate, 3-foliate, leaflets toothed, obovate, rounded, glabrous. **Fl.** White or yellowish-white, axillary, pentamerous. **Fr.** Legume or pods, with persistent beak. **Se.** yellow or greenish-brown.

237. *Triticum aestivum* Linn. Gehun

Family: Poaceae

Herb

Plant

Leaves

Fruits

Buds

Key Characteristics for Identification

H. Tufted herb. **St.** Herbaceous, erect, cylindrical, branched, hollow, smooth, green. **L.** Alternate, sessile, entire, acute, very long and narrow with ligule and auricles. **Fl.** Sessile, complete, spike or 2-ranked compressed spikelets. **Fr.** Cariopsis. **Se.** Whitish grey, ovate or oblong, blunt at both ends, with furrow.

238. *Tylophora indica* (Burm.f.) Merrill.

Family: Asclepiadaceae

Arkapatri

Climber

Plant

Leaves

Fruits

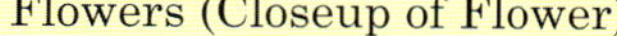

Flowers (Closeup of Flower)

Buds

Key Characteristics for Identification

H. A lactiferous, climber. **St.** Elongated, glabrous or pubescent. **L.** Simple, opposite, somewhat fleshy, ovate or elliptic-oblong, acute or acuminate, often apiculate, glabrous, more or less pubescent beneath. **Fl.** Greenish-yellow, purplish, in umbellate cymes. **Fr.** Folicles, fusiform, divaricate, glabrous. **Se.** Ovate

239. *Typha elephentina* Roxb. Eraka

Family: Typhaceae Shrub

Plant

Leaves

Leaves Cross Section

Fruits Condition 1

Fruits Condition 2

Fruits Condition 3

Seeds

Key Characteristics for Identification

H. A gigantic gregarious marsh herb. **L.** Erect, grass like leaves, convex dorsally and concave ventrally. **Fl.** Flowering stem embraced at the base by the leaf-sheaths, straight, glossy, forming the rhachis of the female and male spikes. Lower female spike, finally brown. Female flowers with lanceolate stigma.

240. *Urginea indica* (Roxb.) Kunth.

Family: Liliaceae

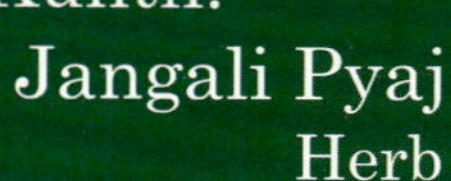

Jangali Pyaj

Herb

Plant

Flower

Leaves and Tubers

Tubers

Key Characteristics for Identification

H. An annual herb. **Bulb:** Pale, ovoid, thick, very bitter. **L.** Radical, flat, linear, acute. **Fl.** Scape erect, flowers greenish-white or dingy-brown, very distant, in slender laxly flowered racemes 15-30 cm long. **Fr.** Capsule, ellipsoid, tapering at both ends, the cells 6-9 seeded, **Se.** Elliptic, 6 by 3 mm, flattened, black.

241. *Vernonia cinerea* Less. Sahadevi

Family: Asteraceae

Herb

Plant

Leaves

Flowers

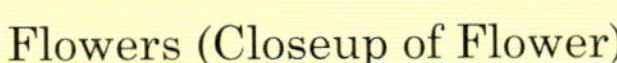

Flowers (Closeup of Flower)

Flowers and Seeds

Key Characteristics for Identification

H. A decumbent herb. **St.** Stiff, cylindric, striate, more or less pubescent, slightly branched. **L.** Simple, alternate, variable in shape, broadly-elliptic or lanceolate, obtuse or acute, shortly mucronate, crenate-serrate. **Fl.** Pinkish-violet, in small heads. **Fr.** Achene, oblong, clothed with white hairs. **Seed** Small.

242. *Vitex negundo* Linn. Nirgundi

Family: Veranaceae Shrub

Plant

Leaves (Inset Reverse Leaf)

Flowers (Inset Closeup Flower)

Inflorescence

Seeds

Key Characteristics for Identification

H. A large shrub or small tree. **St.** Bark thin, grey, branchlets quandrangular. **L.** Opposite, digitately 3-5 foliate, leaflets lanceolate unequal, glabrous, middle one longer. Specific smell in crushed leaves. **Fl.** Bluish- purple, crowed in short cymes. **Fr.** Drupe, sub globose or ovoid, 4-seeded. **Se.** Obovate or oblong.

243. *Vitis vinifera* Linn. Drakhsa

Family: Vitaceae Climber

Plant

Leaves

Unripe Fruits

Ripe Fruits

Dry Fruits

Seeds

Key Characteristics for Identification

H. A large, tendril climber. **St.** Tendrils leaf-opposed, bifid. **L.** Orbicular-cordate, more or less deeply 3-5 lobed, margin irregularly toothed, glabrous. **Fl.** Small, greenish or white, leaf-opposed, penicled cymes. **Fr.** Berry, globose, ovoid or oblong, bluish black, purplish or greenish pale. **Se.** Oblong-obovoid, brown.

244. *Withania somnifera* (L.) Dunal.

Family: Solanaceae

Aswagandha

Herb

Plant

Leaves

Flower and unripe fruits

Fresh Fruits

Ripe and Unripe Fruits

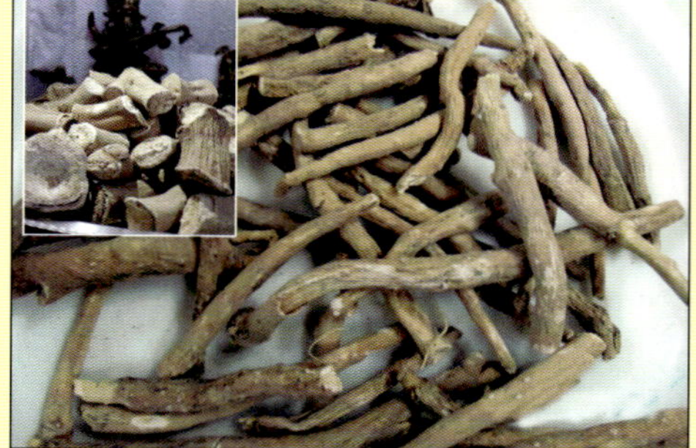

Dry Roots (Inset Dry Roots of Nagaori)

Seeds

Key Characteristics for Identification

H. A undershrub. **Root:** Cylindrical, brownish-white surface. Taste is bitter. **St.** Herbaceous, covered with wooly hairs. **L.** Alternate, ovate, entire, glabrous. **Fl.** Greenish-yellow or lurid yellow, bell-shaped. **Fr.** Berry, globose, orange coloured when matured, enclosed in a persistent calyx. **Se.** Flat, yellow, reniform.

245. *Woodfordia fruticosa* Kurz.

Family: Lythraceae

Dhataki
Shrub

Plant

Leaves

Flower and Leaves

Flowering Branch

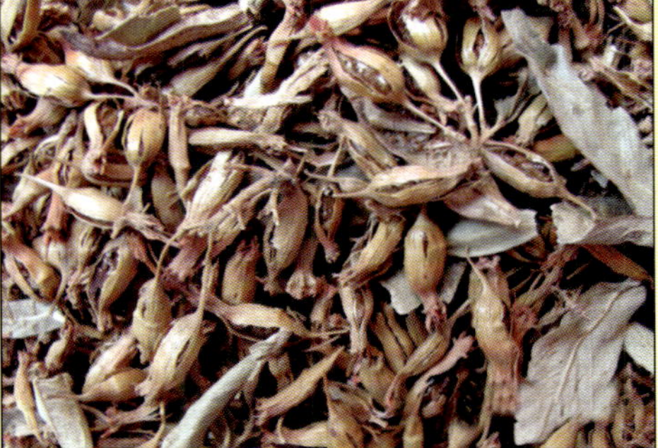

Dry Flowers

Fruits

Key Characteristics for Identification

H. A beautiful, straggling leafy shrub. **St.** With many long arching branches and smooth, brown bark, young shoots terete, young branches quandrangular. **L.** Opposite, ovate-lanceolate, lanceolate, acute. **Fl.** Bright red, in dense 2-15 flowers. **Fr.** Capsule, ellipsoid, membranous. **Se.** Brown, ovate or oboviod.

246. *Wrightia tinctoria* (Roxb.) R.Br.

Family: Apocynaceae

Khirani

Tree

Tree

Leaves (Inset Stem)

Fruits

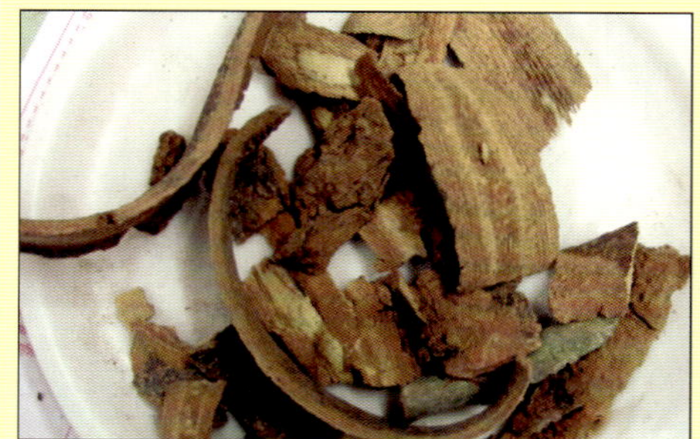

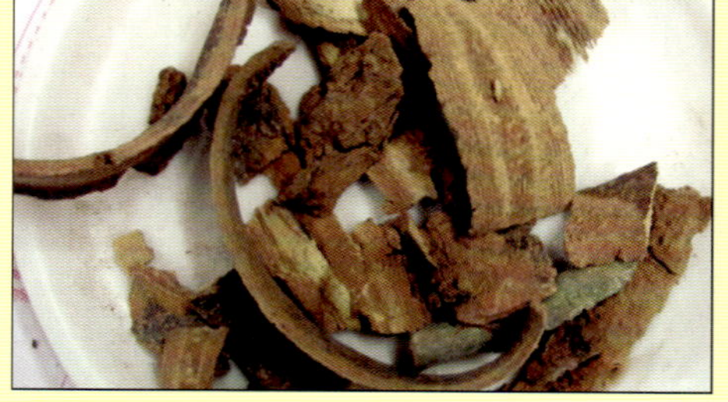

Stem Bark

Flowers (Inset Closeup Flower)

Dry Seeds

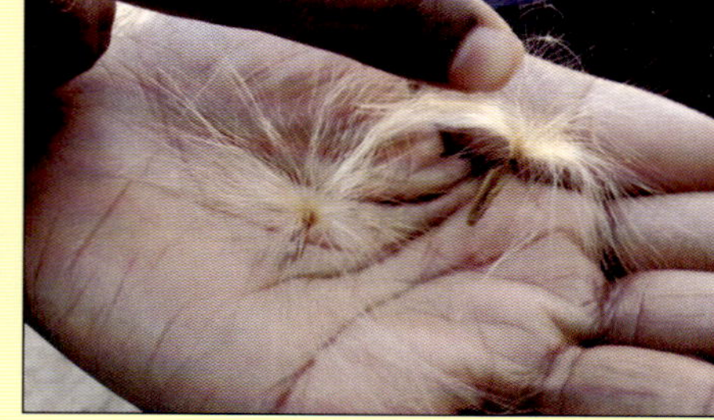

Seeds

Key Characteristics for Identification

H. A small deciduous tree. **St.** Branchlets glabrous or puberlous, with milky latex. **L.** Oblong-lanceolate, acute or acuminate, glabrous or puberlous beneath. **Fl.** White or yellowish, fragrant. **Fr.** Follicles, dropping, united at apex, when young, pendulous, striete. **Se.** Linear-oblong, pale brown, coma twice.

247. *Xanthium strumarium* Linn.

Family: Asteraceae

Aadhashishi
Herb

Plant

Leaves

Ripe Fruits (Closeup of Ripe Fruit)

Unripe Fruits

Key Characteristics for Identification

H. A tall coarse herb. **St.** Short, stout, rough. **L.** Numerous, alternate, trinagular-ovate or suborbicular, acute, often 3-lobed, toothed. **Fl.** Numerous, white or. **Fr.** Achenes, 1.3 cm long, oblong-ovoid, thick, anclosed in the hardened, involucral bracts and clothed with strong hooked spines. **Se.** Numerous.

248. *Zea mays* Linn.

Family: Poaceae

Makka

Herb

Plant

Flower

Fruit

Flower unripe

Seeds

Key Characteristics for Identification

H. A robust herb. **St.** Smooth. **L.** Large, with sheathing base, ciliate ligule and broadly lanceolate blade. **Fl.** Male spiklets in long spike-like racemes, female inflorescence in the axil of the leaves 2-fid at the tip, papillose upwards, surround by the dried up glumes, valves and valvules. **Se.** Large, sub-globose.

249. *Zingiber officinale* Rosc. Shunthi

Family: Zingiberaceae

Herb

Plant

Leaves

Flowers

Fruits

Dry Tuber

Tuber fresh

Key Characteristics for Identification

H. A slender rhizomatous herb. **R.** White to yellowish brown, irregularly branched, somewhat annulated and laterally flattened. **St.** Erect leafy. **L.** Sessile, linear-lanceolate, glabrous, acute or acuminate. **Fl.** Yellowish-green, with a small dark purple or purplish black lip. **Fr.** Capsules, oblong. **Se.** Globose, arillate.

250. *Ziziphus jujuba* (Lam.) Gaertn. non Mill.

Family: Rhamnaceae

Ber Bara

Tree

Tree

Stem

Leaves

Flowers

Fruits

Dry Fruits

Key Characteristics for Identification

H. A large thorny shrub or small tree. **St.** Bark rough grey or black, prickles on stem and branches, in pairs, straight and curved. **L.** Alternate, oblong or ovate, glabrous upper surface, beneath with white tomentum. **Fl.** Greenish-yellow. **Fr.** Drupes, globose or ovoid, fleshy, yellow to orange and red. **Se.** 1-2.

251. *Ziziphus nummularia* (Burm.f.) Wight & Arn.

Family: Rhamnaceae

Ber Chota

Herb

Plants

Leaves

Flowers

Fruits

Root Bark

Dry Fruits

Key Characteristics for Identification

H. A small spinous shrub. **St.** Branches diverciate, slender, zigzag, bark light coloured. **L.** Orbicular or ovate, beneath with a tomentum; thorns in pairs, one straight, sharp other, hooked. **Fl.** Small, greenish-yellow. **Fr.** Drupes globose, glabrous, green unripe, yellow and turning red when ripe. **Se.** 1-2 celled

Index of Ayurvedic/Hindi Names

(Please note that Numbers given after Plant is the Serial Number of Plant, not Page Number)